D0910269

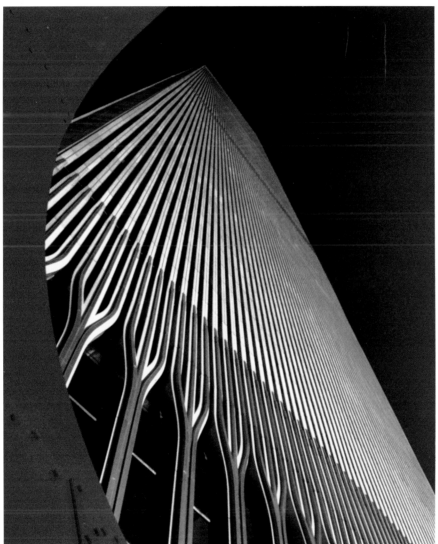

TOWER OF THE WORLD TRADE CENTER, NEW YORK CITY

THE BUILDERS

MARVELS OF ENGINEERING

Published by
The Book Division
National Geographic Society
Washington, D.C.

Published by
The National
Geographic Society

Gilbert M. Grosvenor
**President and
Chairman of the Board**

Michela A. English
Senior Vice President

Prepared by
The Book Division

William R. Gray
Vice President and Director

Margery G. Dunn
Charles Kogod
Assistant Directors

Staff for this Book

Elizabeth L. Newhouse
Editor

Cinda Rose
Art Director

Thomas B. Powell III
David Ross
Illustrations Editors

Carolinda E. Hill
Victoria Garrett Jones
Ann Nottingham Kelsall
Shelley L. Sperry
David J. Whitaker
Researchers

Margery G. Dunn
Edward Lanouette
Contributing Editors

Lyle Rosbotham
Contributing Art Director

Leslie Allen
Mary B. Dickinson
Dónal Kevin Gordon
Carolinda E. Hill
Catherine Herbert Howell
Alison Kahn
Ann Nottingham Kelsall
Edward Lanouette
Jane L. Matteson
John F. Ross
Anne E. Withers
Writers

Sandra F. Lotterman
Editorial Assistant

Artemis S. Lampathakis
Illustrations Assistant

Karen F. Edwards
Design Assistant

Richard S. Wain
Production Project Manager

Lewis R. Bassford
Heather Guwang
H. Robert Morrison
Production

Elizabeth G. Jevons
Teresita Cóquia Sison
Marilyn J. Williams
Staff Assistants

Bryan K. Knedler
Indexer

**Manufacturing and
Quality Management**

George V. White
Director

John T. Dunn
Associate Director

R. Gary Colbert
Executive Assistant

Diagrams by
Dale Glasgow

Joyce B. Marshall
Elizabeth P. Schleichert
Contributors

Copyright © 1992 National
Geographic Society. All rights
reserved. Reproduction of
the whole or any part of the
contents without written
permission is prohibited.
Library of Congress CIP data:
Page 288.

Foreword

Tower and cables come together on California's Golden Gate Bridge, a 20th-century landmark that combines technology and art.

Building to withstand nature's mighty powers—storms, floods, earthquakes, and gravity—engineers serve us well. Engineering evolved over 5,000 years, from piled stones to steel skyscrapers, from primitive rope bridges to the immense spans serving cities worldwide. The story combines progressively more effective uses of earth resources such as clay, stone, and iron with modern mathematics and machines and new materials. Often the results are beautiful, soaring forms or immense, abstract sculptures, as engineers readily cross the line between utility and art. "Engineer" and "ingenuity" come from the same Latin root.

Roads, dams, canals, and tunnels are purely functional. But all structures, whatever their purpose, must carry weight from above, and most need defenses against wind, earthquakes, or water pressure, as well as against the destabilizing stresses that monumental works carry within themselves. Thus, Gothic cathedral vaults are buttressed, and the U.S. Capitol dome is built over a metal frame. Engineering know-how is always needed.

As construction proceeds, materials must be brought to the right place at the right time, not just for efficiency's sake but to maintain stability and balance while work goes forward. Sandhogs at a tunnel head or wire-spinners high above a suspension-bridge roadway work in carefully choreographed sequences specified by engineers. Exact organization exists also within all successful designs, to equalize and pass on forces acting on the structures, as the weight of a stone bridge's upper part is carried down to bedrock by a row of arches that stabilize each other.

Early pyramid specialists mastered quarrying, stonecutting, and the mechanics of moving, hoisting, and placing huge stones. Calculations were secret, passed from masters to apprentices. Architect-engineers in Greece and Rome were ranked high and were expected to plan cities, build aqueducts and irrigation facilities, and design roads, bridges, ports, and fortifications. In Asia, engineers were civil servants; in Europe, they came from military and ecclesiastical backgrounds.

Simple wood and masonry construction remained the rule until the 1700s, when scientific advances and new urban and commercial requirements caused far-reaching changes. What had been made of wood, stone, or bricks and mortar now rose in iron, steel, and glass. New methods made it possible to build structures of unheard-of size and complexity to serve the many needs of large populations.

Successful engineering requires strong foundations and structures that give somewhat in gales and earthquakes. Flexibility is built in by using relatively small building elements, fastened one to another. London's Crystal Palace of 1851 forecast modern designs: Its pieces of wood, metal, and glass, largely prefabricated and cut to standard sizes, simplified construction and formed a strong, lightweight, and transparent shell. Today, advanced research permits near-miracles of skyscraper tube construction, new kinds of river and sea barriers, and the enormous unobstructed interiors of sports arenas like those of New Orleans and Houston.

Great engineering achievements evoke strong responses because they give us something we never had before. Many are great monuments, strong and solid against the sky; others, underground, are no less remarkable. Every one is a crucial factor in the web of civilized life; every one assists in making something essential happen. Clarity of design makes identification easy and can create international symbols—the Eiffel Tower, the Golden Gate Bridge. That such structures often are handsome stems both from elementary, irreducible shapes and from the inherent attractiveness of logical forms and their mathematical basis.

Engineering is essentially a social art, increasing in importance as life becomes more complicated. We depend on it as we do on nature itself, for it makes nature more tractable and accessible.

William L. MacDonald, *Building Historian*

OVERCOMING DISTANCE

OVERCOMING DISTANCE

ROADS
CANALS
BRIDGES
RAILROADS
PIPELINES

They are at once the lifelines of modern civilization and the means for its dispersal: roads and highways that carry trucks and automobiles; canals and inland waterways that ship commodities near and far; bridges that leap rivers, canyons, and other obstacles; railroads that move freight; and pipelines that transport water, natural gas, and petroleum and its liquid byproducts.

For modern civil engineers, the unimpeded movement of goods and people is among the most fundamental objectives, one that embraces a host of challenges—from the desire to create works that are efficient, pleasing, and economical to the need for altering the landscape.

Building large structures invariably requires excavation, and earth moving, even with today's arsenal of equipment, is not easy. Hills must be resculptured to accommodate a highway, a river diverted while a dam is constructed. Some structures even permanently transform the earth's topography, connecting the Pacific Ocean with the Caribbean Sea, as the Panama Canal did in 1914, or opening a new artery into the heart of North America, as the St. Lawrence Seaway accomplished in 1959.

But rearranging the topography is only one step in the construction process. It begins with an investigation of the proposed site, during which routes are surveyed and mapped. Engineers do test

THE GEORGE WASHINGTON BRIDGE,
NEW YORK AND NEW JERSEY

ROMAN AQUEDUCT, SEGOVIA, SPAIN

INTERSTATE 25, DENVER, COLORADO

drillings and use the resulting core samples to determine whether the soil along a proposed road or pipeline is stable enough to support the structure, or whether underlying bedrock can hold the foundation of a bridge. Still later in the process, building materials are chosen after climate, site selection, function, appearance, economics, and other factors have been calculated. Computers increasingly assist in analysis, as well as in the overall design.

PRECEDING PAGES: Defying the time-honored theorem that the shortest distance from here to there is a straight line, the Trans-Alaska Pipeline zigzags for miles across snow-covered tundra, riding a "high-rise ditch" composed of 78,000 vertical supports sunk deep into the frozen permafrost.

The structure's zigzag path gives it leeway to expand and contract safely as temperatures rise and fall and to survive in the event of an earthquake. Anchoring the pipeline on special platforms at its elbows enhances its structural stability.

In addition, a range of other criteria must be taken into account, depending on the type of structure. In designing a road, for example, engineers must consider the volume of traffic, the kinds of vehicles that will use it, and how fast the vehicles travel.

Similar factors play a role in bridge design; the requirements for railroad and automobile bridges differ. The length of the span is one essential element in determining the design. Engineers must also consider the effects of wind and weather and the impact of tides and currents on underwater foundations. Moreover, with bridges aesthetics play a part: The best bridge engineers of the past two hundred years have also been structural artists, making the most of new materials—from iron to prestressed concrete—to create works of clarity, efficiency, and elegance.

In waterway design, channels must have either moderate gradients or locks to move vessels from one water level to another; bank erosion is a concern. Railroads must be designed with limited gradients. Also, the track and roadbed must withstand the weight of passing trains, as well as the dynamic loads imposed by the impact and sway of speeding locomotives.

Major works of civil engineering, today usually undertaken by governments, are hallmarks of civilization. Often, like the first transcontinental railroad and the Erie Canal, they play vital roles in a nation's or a region's development; like the Roman aqueducts and the Garabit Viaduct, the George Washington and the Ganter Bridges, they can radically alter the look of the landscape. Well designed and built, they serve as symbols of their eras and evidence of human mastery over seemingly insurmountable challenges.

THE TRANS-ALASKA PIPELINE

THE CANADIAN PACIFIC RAILROAD, ALBERTA

Roads

Symmetrical ramps and a multilane overpass mark an interchange of the 750-mile Los Angeles freeway system.

Begun in 1811 and completed some 40 years later, the National Road stretched more than 600 miles between Cumberland, Maryland, and Vandalia, Illinois, and helped open up the country to western settlement. It got its start as an old Indian trail, later widened for wagons. Near Washington, Pennsylvania, in 1910, its original surface shows the wear and tear of automobile traffic, but forms a solid base for future asphalt paving. This was the first U.S. highway constructed entirely with federal funds. Congress wanted not only the straightest route, but also one that never exceeded a grade of five degrees. A compromise was reached: gentle curves around mountains, then straight ahead; the grade requirement was almost always met. Earlier sections, consisting of a sand or gravel surface over two layers of stones, took a beating from heavy wagon and coach traffic—and droves of cattle. Later ones were wider and macadamized. Today, the National Road lives on as part of U.S. Route 40.

The ancient Romans built roads to last; some of them, like parts of the Appian Way, have already endured for more than 2,000 years. To the Romans we owe the term "highway." They recognized the importance of proper drainage. When digging the roadbed, laborers heaped up earth from parallel ditches. This elevated the road above the surrounding terrain—creating, literally, a high way.

The Romans were also no-nonsense road builders, especially when it came to the first-class highways maintained by the state—the *viae publicae,* or public ways. Not for them any curves or dips and rises. Major Roman highways usually were arrow straight from point to point. Even without dynamite and bulldozers, the Romans were masters of cut and fill—slicing through hills, cracking rocks by heating them with fire and then dashing them with cold water, spanning ravines with arched bridges, and elevating low spots with earthen fill.

Major Roman highways ranged up to 30 feet wide and carried two lanes of traffic. They were narrower in mountainous or other difficult terrain. A surveyor laid them out, usually by sighting point to point and driving stakes into the ground to mark the route. Cross-staffs hung with plumb bobs helped the builders check the level and course of the road.

14

Roads

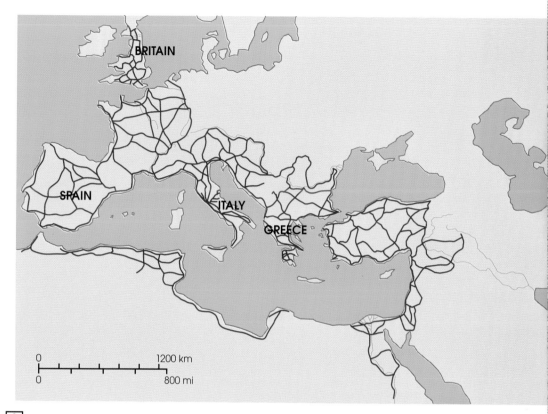

S oldiers, slaves, and convicts provided much of the muscle needed to build a Roman road. First they cleared brush, then dug drainage ditches and leveled the roadbed. Heavy rollers hauled by hand compacted the subsoil. Then layers of stones were set into concrete and tamped with weights fastened to poles. Finally, thick stone slabs were fitted into place to form a strong, load-bearing surface edged with curbstones. A major road might be up to five feet thick and was cambered (arched slightly in the center) to shed rainwater.

Over such roads freight hauled in

A Roman road near Aleppo, Syria, still carries traffic after two millennia. A 50,000-mile network (above) linked the far-flung empire.

Cutaway drawings reveal the layers of a Roman road traversing dry ground and a marshy area. Concrete helped hold the road together.

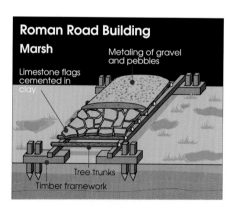

Roman Road Building

Marsh

Metaling of gravel and pebbles

Limestone flags cemented in clay

Tree trunks

Timber framework

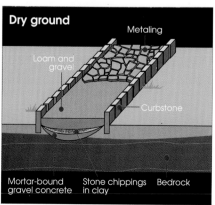

Dry ground

Metaling

Loam and gravel

Curbstone

Mortar-bound gravel concrete · Stone chippings in clay · Bedrock

Roads

The Inca Trail scales the heights to Machu Picchu in Peru. With neither wheel nor draft animal, the Inca built a 10,000-mile network of roads with a "labor tax" imposed on villagers.

"It is more difficult to go to Sichuan than to get into heaven," remarked an eighth-century Chinese poet. The winding road at right, built in the 1960s, extends more than 1,400 miles across Sichuan and Tibet—a two-week drive.

In the 1980s, an army of Chinese peasants pitches in to widen a remote, 30-mile stretch of road. Employing as many as 200,000 manual laborers, the project took only a week.

A villager in the Peruvian Andes shepherds her llamas along an Inca road. Such routes included both earthen and stone-paved walkways.

carts or wagons moved 15 to 75 miles a day, and couriers on horseback, working in relays, could race nearly 150 miles in 24 hours, pausing at way stations for food and fresh horses.

Roman road building began almost full-blown in 312 B.C., when the senate approved construction of a 132-mile highway from Rome south to Capua—the Appian Way, named for its builder, Appius Claudius. As the empire expanded over the next six centuries, the Via Appia became part of a system of some 50,000 miles linking Rome with its outposts in Europe, the Middle East, and Africa.

Although the Romans were by far the best road builders the world had yet seen, they were by no means the first. The Persians and the Chinese had roads as long ago as 500 B.C., but most of them were primitive by Roman standards, and lack of maintenance inspired a Chinese saying that a road was good for ten years and bad for ten thousand. With the collapse of the Roman Empire in the fifth century A.D., road building became a lost art, and Rome's once magnificent highway system gradually fell to ruin.

The modern road traces its beginnings to Pierre Trésaguet, a French engineer who in 1775 developed a new kind of

R o a d s

road—much lighter than the massive, solid Roman ones. His roads were 18 feet wide and only 10 inches thick. To build them, he laid fieldstones edgewise and covered them with crushed and graded rock. Thus the subsoil, rather than thick stone slabs favored by the Romans, bore the weight of traffic. Such a road was far easier and more economical to build—and required much less moving of earth.

Trésaguet's counterpart in England was John Metcalf—known as Blind Jack of Knaresborough because a childhood bout of smallpox had left him sightless. He, too, believed in lighter roads supported by properly drained subsoils. So successful were his efforts that he built some 180 miles of roads and bridges. Through one marshy section, he "floated" his road on a thick subbase of gorse and heather laid crisscross to form a firm foundation.

Building on Metcalf's and Trésaguet's technique, two other English pioneers, Thomas Telford and John McAdam, in the 1820s and '30s perfected the use of stone sorted by size into layers and compacted into a watertight surface by passing wagon wheels. Both men insisted on adequate drainage and cambered their roads.

To achieve adequate drainage, Telford cut ditches and elevated his roads three or four feet above ground level. Large foundation stones, with their flat surfaces down, were placed by hand. They were then covered with smaller stones, which carriage traffic compacted.

McAdam, recognizing the load-bearing capacity of dry soil, also raised his roads. He built them up with several layers of rough stones, each layer decreasing in size toward the surface. Passing traffic pul-

John Loudon McAdam, improver of Britain's roads, was born in Scotland in 1756. Unlike other engineers of the day who designed roads with large foundation stones, McAdam believed that the soil itself could bear the weight of traffic. To facilitate drainage, he elevated his cambered roads and dug ditches along both sides. Several layers of stones, naturally compacted by traffic, made up the macadamized roads widely found in Britain and the United States until the early 1900s.

verized and compacted the stones into a water-resistant surface. Telford's and McAdam's works became standards of excellence until the coming of the automobile and the inflatable rubber tire around 1895.

Rubber tires, unlike slower moving, iron-rimmed wagon wheels, grabbed and spewed grit and gravel into the air, quickly

The Pan American Highway, a cooperative venture originally conceived in 1923, climbs through hills near Quetzaltenango, Guatemala. The United States provided plans, specifications, and construction help for parts of this 30,000-mile road network linking the Americas.

Roads

wearing away macadamized roads. Something was needed to provide a more durable surface. As early as 1810, French and English road builders had been experimenting with natural tars and various cements mixed with stones and pebbles. In 1824, Joseph Aspdin concocted "Portland cement" by burning a mixture of clay and lime. He had rediscovered the art of cement making known to the Romans. In 1865, the mixture was first used to pave a road in Scotland. Its hard, waterproof surface made an ideal paving material and became a forerunner of the modern concrete highway.

Meanwhile, in 1837, it was found that natural asphalt could be powdered when hot, and then the powder could be compacted by roller into a smooth, watertight

A motorist stops to pay a toll on the Pennsylvania Turnpike—America's first superhighway, opened to traffic in October 1940.

The Trans-Canada Highway threads through Rogers Pass in British Columbia (opposite).

Named after a poem by a ninth-century Buddhist monk and famed for its views, Japan's Irohazaka Driveway has 48 hairpin curves.

pavement. A year later, two French engineers invented a heavy steamroller.

In the United States, asphalt came into use in 1877, when Amzi Lorenzo Barber acquired the rights to extract pitch from a huge black lake in the West Indies. Soon he had a contract to pave Pennsylvania Avenue, then a rutted, dusty boulevard in the nation's capital. So successful was this experiment that, by the late 1800s, Barber's Trinidad Asphalt Company had paved the streets of Buffalo, Chicago, New York, and San Francisco.

Building equipment kept pace with the spread of roads, evolving from primitive horse- or mule-drawn scrapers in the late 1800s to today's giant, diesel-driven earthmovers and paving machines that can lay down a ribbon of asphalt or concrete at a rate of 30 feet or so a minute.

New York's Bronx River Parkway, the forerunner of the limited-access expressway, was opened in 1923, followed by a parkway system in nearby Westchester County and on Long Island. Germany and Italy, too, began their respective autobahn and autostrada systems in the 1920s.

By the end of World War II, it was clear that most of America's roads and streets were inadequate to the demands of modern traffic. Many roads were underdesigned and obsolete—too curvy, too narrow, too weak for heavy loads. Bridges weren't sturdy enough; tunnels and overpasses were low. A massive interstate highway building program was begun in 1956. Now nearing completion, it will link every major city in the United States in a 50,000-mile network of multilane, divided, high-capacity highways—a project as ambitious as that started by Appius Claudius more than 2,300 years ago.

I-70 Glenwood Canyon

The Colorado state legislature was emphatic: The new stretch of highway through Glenwood Canyon west of the Continental Divide should be "so designed that . . . the wonder of human engineering will be tastefully blended with the wonders of nature."

The new section, scheduled for completion in 1993, will be part of Interstate 70, a transcontinental highway reaching from Baltimore, Maryland, to Interstate 15 south of Salt Lake City, Utah; it will close Interstate 70's final, 12.5-mile gap. Getting the job done will have involved formidable and sometimes unique challenges.

Glenwood Canyon, carved by the fast-flowing Colorado River over the last 40 million years or so, plunges more than 2,000 feet at its deepest—and is very narrow at the bottom. Although scenic, the canyon is not a wilderness. Early trappers and settlers hacked a wagon trail along the north bank, and in 1887 railroad tracks were laid on the south side. The first cars—a pair of Wintons—drove through in 1902. Since then, a hydroelectric dam, a sheep ranch, and a narrow, two-lane highway—U.S. Route 6, built in the 1930s—have shared the canyon's rocky depths.

The antiquated roadway had become a dangerous bottleneck to east-west travel in this part of the state. The problem was how to upgrade it to interstate highway standards without also

destroying a delicate environment. Normal cut-and-fill road-building methods would irreparably damage the canyon. And the highway would have to be kept open to regular traffic during construction; a handy detour simply didn't exist.

Recognizing the delicacy of the situation, the Colorado Department of Transportation (CDOT) early on sought to involve local citizens' groups. It also hired "inspired" designers from outside the state: Joseph Passonneau of Washington, D.C., and Edgardo Contini of Los Angeles. The success of the project would depend not only on their designs, but also on their ability to win the confidence of opponents— of whom there were many.

Lawsuits, environmental impact studies, and design changes took nearly two decades to resolve before the first shovelful of earth was turned in 1980. Bitter opposition gradually gave way to overwhelming approval. Ralph Trapani, project manager for the CDOT, put it this way: "Old, grouchy highway engineers [wound up arguing] to save a tree. Then the environmentalists would talk safety. It was the most complete flip-flop of culture I've ever seen."

Working with citizens, the engineers designed not only an extraordinary highway, but also a ten-foot-wide bike path, rest areas, and riverside boat launches.

Railroad tracks and a two-lane road built in the 1930s bracket the Colorado River in a picture taken before construction began. In this narrow, winding canyon, the major challenge lay in building a safe, efficient highway without degrading the environment.

Roped like mountaineers, workers (opposite, upper) use pry bars to remove potentially hazardous loose rock and debris from the face of a cliff after blasting operations.

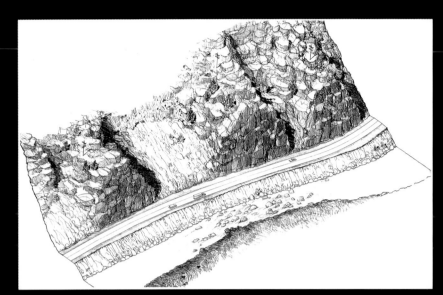

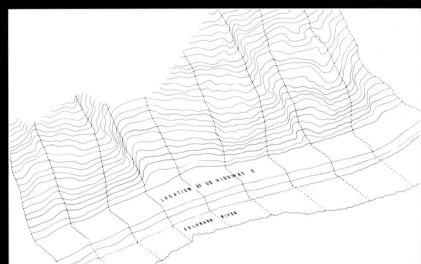

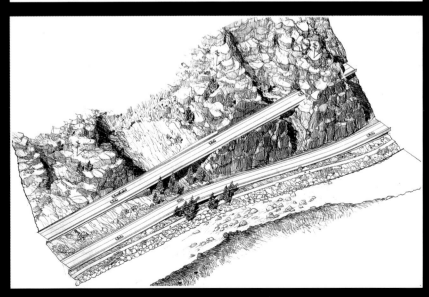

Isometric projections, derived from feeding data into a computer, help engineers visualize highway alignments. Using the projection of a section of Glenwood Canyon (middle illustration) as a base, an artist sketched an isometric drawing of the old road (top) and then one of the new road as it would appear (bottom).

Swirling steel propped on concrete piers (upper) shapes the Shoshone Dam interchange near the center of Glenwood Canyon. When completed, the ramp will provide access to the dam and to the Hanging Lake rest area.

Temporary steel columns steady the Amphitheater Bridge during construction. The brown, sling-like cloth prevents epoxy that glues the bridge segments together from dripping into the river. The traveling gantry sits on the completed section of the roadway.

As it does for all interstate highway projects, the federal government picked up 90 percent of the tab—in this case, half a billion dollars. The Federal Highway Administration agreed to ease certain design standards normally applicable to the interstate system. Tight spaces and a tender environment at times demanded sharper curves and narrower-than-normal shoulders.

The solution was to build a terraced, four-lane highway, with westbound lanes stepped above those heading east. In places the roadbeds would extend six feet out over reinforced retaining walls so that deep shadows and shrubs planted along the footings would help conceal the walls. The bike path, separated from the din of traffic, would run the length of the canyon between the river and the road. And erosional scars left from the construction of the old highway would be regraded and replanted with native shrubs and trees.

There were rules for contractors, too. Those who needlessly destroyed vegetation were subject to fines—$30 for a raspberry bush, $45 for a scrub oak, more than $20,000 for a mature blue spruce or cottonwood. Fresh rock exposed by blasting had to be chiseled into cragginess and stained to blend in with weathered rock.

The canyon's steepness and narrowness required that much of the roadway be elevated on 39 bridges and viaducts with a combined length of more than six miles. These structures are supported on slender columns, or piers, painted to resemble the natural rock, and some of them soar as high as 80 feet above the ground.

Lack of elbow room at construction sites made it necessary to truck in many prefabricated road and bridge segments.

I-70 Glenwood Canyon

At some bridge sites there wasn't even room for a regular crane to hoist pieces into place without disrupting traffic or disturbing the landscape. Contractors solved the problem by importing a remarkable French gantry crane similar to those used to build bridges in the Alps. Its articulated, 350-foot boom, mounted on supporting "legs," travels along completed sections, maneuvering additional segments into place from above.

To avoid intruding on a popular scenic attraction, the trail to Hanging Lake, the highway sweeps across the river on long spans and disappears into twin 4,000-foot-long tunnels. Because of the length and isolation of the tunnels, air quality and traffic are monitored from a four-story underground control center nearby.

Closed-circuit television and sensors embedded in the bridge decks relay information about ice and other hazards to the control center so that maintenance crews can respond anywhere along this elegant, high-tech highway.

At work on the French Creek Viaduct (upper, left), the gantry's 350-foot boom lowers a 40-ton precast bridge segment into place.

Workers on a suspended platform (upper, right) cut steel rods to length. The tensioned rods help make the bridge segments secure.

Supported on piers, a new roadway rises above the old as both bend around a cliff along the Colorado. Here engineers provided a bench for the highway by cutting into the canyon wall. Now open, the upper section carries westbound traffic.

25

Canals

France's Languedoc Canal, opened in 1681, extends nearly 150 miles between Toulouse and the Mediterranean Sea.

ROME'S MULTILEVEL AQUEDUCTS CARRYING WATER TO THE IMPERIAL CITY.

For thousands of years, one of the greatest engineering challenges has been to bring water where it is needed, whether to irrigate crops or create shipping routes. During the Renaissance in Europe—and much earlier in China, Mesopotamia, and Egypt—builders cut channels through the land and erected dams to hold water.

The Romans built bridgelike aqueducts. Canal locks were probably invented by the Dutch in the 14th century, enabling vessels to cross hills and detour around falls.

By the mid-19th century, European canal engineers were slicing through land barriers, shortening world shipping lanes; the Suez Canal still serves as a maritime shortcut. Today, the St. Lawrence Seaway draws oceangoing ships into the heart of North America, offering economical transport for commodities.

Over the years people have learned to harness gravity to move water and goods. They have also learned to overcome gravity with lifting devices that range from ancient waterwheels to modern hydraulic pumps that can sluice thousands of gallons of water a minute.

The Grand Canal

M ost of China's major rivers run from west to east. As early as the fourth century B.C., Chinese emperors realized that if they built north-south canals to link the river systems, they could unify a vast empire—and ensure a steady supply of grain from the agricultural south to their northern capitals. In stages a great waterway grew across China, built by multitudes of peasants. Under the harsh rule of early seventh-century A.D. Sui emperors, 5.5 million laborers are said to have worked six years to construct a 1,500-mile section of the canal between the port of Hangzhou and the capital at Luoyang. Two million of them were "lost."

By later in the seventh century, the Tang and Northern Song dynasties were moving north on barges more than

A Chinese dragon of barges weaves along the Grand Canal near the ancient city of Suzhou, which became rich from canal-borne commerce—and known as the home of poets.

300,000 tons of grain a year, along with paper and luxury goods, over a waterway system known as the Grand Canal.

One of the first true summit canals, the Grand Canal follows the contour of the land. Though the terrain is mostly flat, a slight gradient accumulates over its length; to avoid undesirable currents, stone and timber weirs were placed every three miles. Feeder canals drew water from distant rivers into reservoirs, and sluice gates regulated water levels. One hilly stretch required 60 gates. Where a rise proved too abrupt, boats were hauled up slipways.

Parts of the canal have remained navigable for 2,000 years. Repaired and redredged by Kublai Khan in the 13th century and by the People's Republic in the late 20th, it still serves as a route for local commerce. Recently, high-powered pumps have made it part of a flood-control and irrigation system.

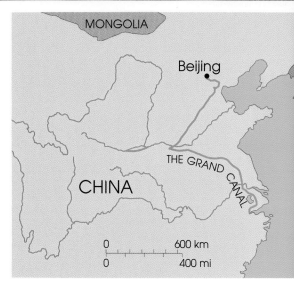

Built to carry grain and other commodities from the south to the northern capitals, the canal still moves goods through eastern China.

A solitary barge motors past Wuxi (top, right), a major grain shipping center along the Grand Canal since the seventh century.

Heavy traffic on 40-yard-wide stretches, such as one in Suzhou (top, left), forces barges to line up end to end—and, often, side by side.

British Canals

Once asked in Parliament what he thought rivers were for, James Brindley (1716-1772) replied, "To feed navigable canals." In devoting his life to designing canals, Brindley changed the face of England. He built the country's first major canal in 1761 and went on to plan a network that totaled 360 miles. Unlettered, he took to his bed to think through a project, then built without written plans. His engineering solutions were as direct as he was. Even when overwork finally felled him, canal promoters stood by his deathbed, seeking last-minute advice.

Curving through farmland near Napton on the Hill, a section of the Oxford Canal has changed little since the 18th century. Rather than build locks as he did here, James Brindley preferred constructing canals on the level. When hills proved too extensive to skirt, he usually tunneled through. During the canal boom of the late 1700s, few places in England lay farther than 15 miles from a navigable waterway.

Britain's industrial revolution got a boost the day the Duke of Bridgewater engaged James Brindley to build a canal leading from his lordship's coal mines at Worsley to the rapidly growing factory town of Manchester. Opened in 1761, the Bridgewater Canal halved the price of Manchester coal and gave the town a head start toward becoming England's premier industrial center. A spate of canal building followed—and continued until the middle of the 19th century.

Britain's hilly terrain presented special problems. Locks—canal enclosures with gates at either end to raise or lower water levels—could stair-step hills, but builders at first avoided them because they were costly to build. Brindley himself designed aqueducts to cross rivers and valleys, and high, solid embankments to carry his canals across dips and swales.

By the early 1800s interlocking canals carried raw materials to burgeoning factories and finished goods to market. Scottish engineer Thomas Telford was completing the 103-mile Ellesmere Canal in western England and Wales, and the Grand Trunk—a network built at the urging of potter Josiah Wedgwood—linked the industrial Midlands with major seaports.

As engineers gained skill, their canals grew straighter—and bigger. Locks came into widespread use. By the late 18th century, engineers could slice their canals straight as a rule, cutting into the Welsh mountains in search of more coal. In the 1840s, England's waterway system encompassed some 5,000 miles of canals and navigable rivers.

Then came a speedier form of transportation that would overtake and eclipse canals—the railroad.

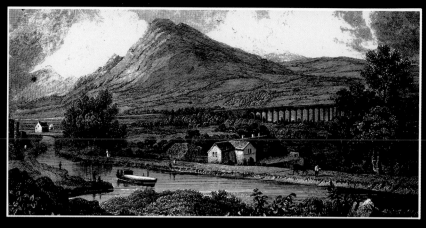

Soaring 120 feet above the River Dee in Wales, Pontcysyllte Aqueduct stands as the single most impressive monument to the skill and daring of late-18th-century British canal builders. Erected by Thomas Telford, a young Scottish engineer, and by William Jessop, chief engineer on the Ellesmere Canal, the aqueduct took advantage of the latest technology—in this case a trough of cast iron, rather than heavy masonry—to form its bed. Iron's much lighter weight enabled Telford to carry the aqueduct on 18 masonry piers that were much slenderer than would otherwise have been possible. Springing from steep embankments, Pontcysyllte Aqueduct reflects Telford's passion for "adventurous structures"—and still gives modern canal cruisers (below) a thrill as they drift above the valley.

The Erie Canal

ANew York surveyor first proposed a canal from the Hudson River to Lake Erie in 1724. Almost a century later, New York politician De Witt Clinton took up the cause and, in early 1817, won both passage of a canal bill and the governorship of New York. By summer, construction on the 363-mile waterway was finally begun.

American canals had been relatively late in coming, and borrowed heavily from Dutch, French, and especially British technology when they did. George Washington was an early promoter of canals, helping to create the Patowmack Canal in Virginia, one of the country's first major ones. Nothing on the scale of the Erie, however, had ever before been tried in the United States.

Eight years after breaking ground, the Erie Canal had become navigable between Albany and Buffalo—and was already helping to shape the nation. Immigrants and goods could move easily through the Appalachians and into the

Great Lakes region—on barges hauled by horses and mules traveling on 10-foot-wide towpaths. New York City had become the country's foremost port, and canalside towns quickly prospered.

The canal engineers, none of them more than surveyors when the project began, trained on the job. On the proposed route, two major hurdles faced them: the 419-foot drop between Utica and the Hudson to the east, and the 60-foot climb up the Niagara Escarpment to the west. Prudently, the engineers set out to gain experience on the relatively flat middle section between the Seneca River and Utica.

This part of the route, through virgin

Pleasure boaters navigate a quiet stretch of the Erie Canal near Middleport, New York, where settlers once barged to the wilderness.

When the canal opened in 1825, Lockport (above)—where five sets of double locks lifted boats 60 feet—became a celebrated town.

The Erie Canal

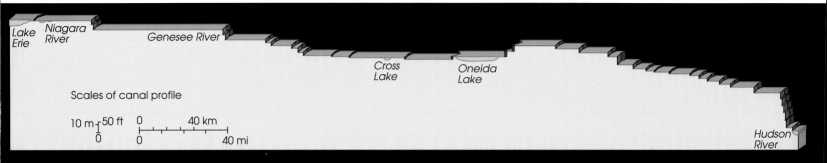

Lake
Erie

Niagara
River

Genesee River

Cross
Lake

Oneida
Lake

Scales of canal profile

10 m ┬ 50 ft   0   40 km
     ┴ 0   └─┴─┴─┴─┘
        0   40 mi

Hudson
River

Climbing 169 feet up from the Hudson River near Waterford, the Erie Canal (below) skirts Cohoes Falls as it begins a westward swing.

Parts of it restored today, the canal (above) extends 348 miles and has 34 locks; the original canal had 83 locks and spanned 363 miles.

forest, required ingenuity. Contractors invented a crank-and-roller device that pulled down trees. Another device, an ox- or horse-powered stump puller, cleared 30 to 40 stumps a day, even green ones. Iron-tipped plows cut the 40-foot-wide, 4-foot-deep channel. The spade and bucket, widely used in Britain, were employed only on wet ground on the Erie.

After the middle section was opened to navigation in 1820, work began on the ends of the canal. In the east, the descent to Albany was challenge enough, but 86 miles of the canal's line also had to be laid through the rugged and narrow Mohawk River Valley. This required the construction of locks and aqueducts, and lining parts of the river itself with masonry. One aqueduct, near Schenectady, spanned 748 feet; another, near Cohoes Falls, 1,188 feet.

In the west, meanwhile, the Irondequoit embankment near Rochester joined

When Benjamin Wright (1770-1842) was named chief engineer of the Erie Canal in 1817, he had had little practical experience in canal building. But by the time he finished the job, he would be acknowledged dean of American civil engineering. At one time a judge whose duties included boundary surveying, Wright himself had helped survey the route. When the project began, Wright quickly realized that since he and his staff were all novices, the canal had to become a school of engineering. He set high standards and cultivated his eye for talent. The result: Until mid-century, almost every civil engineer in the U.S. had trained or been trained by someone who worked under Wright on the Erie Canal.

a series of ridges to raise the canal some 70 feet, and the Genesee River aqueduct carried the canal high above swift currents.

The most dramatic part of the western section, the high climb west of Lockport, was saved for last. In its two-mile Deep Cut, workers had to blast and haul

away 1.5 million cubic yards of solid rock to build five stair-stepping double locks.

In 1825, Governor Clinton inaugurated the canal with a nine-day boat trip from Buffalo to New York Harbor. Amid great fanfare, he emptied a cask of Lake Erie water into the Atlantic Ocean.

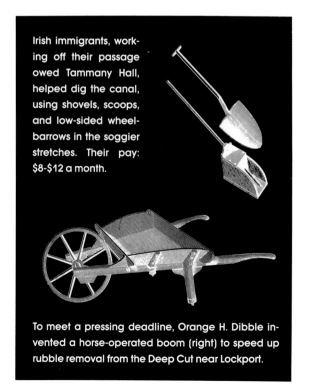

Irish immigrants, working off their passage owed Tammany Hall, helped dig the canal, using shovels, scoops, and low-sided wheelbarrows in the soggier stretches. Their pay: $8-$12 a month.

To meet a pressing deadline, Orange H. Dibble invented a horse-operated boom (right) to speed up rubble removal from the Deep Cut near Lockport.

The Panama Canal

When the Panama Canal opened in 1914, it cut 9,000 miles from a sea voyage between New York and San Francisco. But the long-sought goal of linking the Atlantic and Pacific Oceans took more than 30 years, thousands of lives, 387 million dollars—and ultimately depended on controlling the anopheles mosquito.

Ferdinand de Lesseps, the French entrepreneur who had built the Suez Canal, in 1882 started digging in Panama with French government backing. Nine years later he was bankrupt and had lost some 20,000 workers to cholera, yellow fever, malaria, and other diseases.

President Theodore Roosevelt, aware that Army doctors had pinpointed the mosquito as a transmitter of yellow fever and malaria, seized the chance for the United States to finish the canal.

Cargo ships from different nations meet at the Miraflores Locks on the Pacific end of the canal. Here, ships are lowered or raised 54 feet from sea level. Electric trains act as tugs.

A cruise through Panama, between the Caribbean Sea and Pacific Ocean, avoids a trip around South America—and windy Cape Horn.

Building

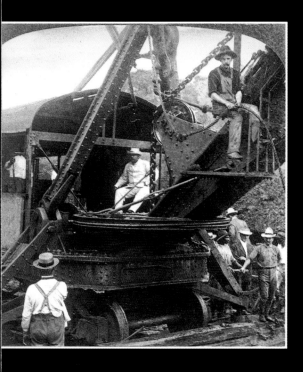

Gargantuan gates at Gatun Locks dwarf their builders in this early 1900s photograph. Riveters attach steel plates to a skeleton of steel girders, making each 700-ton leaf hollow and buoyant—and thus reducing stress on the hinges. Half-submerged, the gates today (left) swing open a crack for an inspector.

Canal booster Theodore Roosevelt (above, left) operates a steam shovel at Culebra Cut, in a hand-colored photograph from 1906.

Every day for two years, crews poured 3,000 cubic yards of concrete to build the Gatun Locks on the Atlantic side (opposite). Rails delivered fresh mix; an overhead cableway lifted and placed it. Despite the feverish pace, the walls have never cracked. A modern-day worker (opposite, far right) crosses the huge floor.

Rather than follow the French plan to cut through at sea level, American engineers—led first by John F. Stevens and later by George Washington Goethals—used locks to raise and then lower the waterway 85 feet. The move meant far less digging—fortunate in light of the unstable earth at nine-mile-long Culebra Cut, where the canal had to cross the Continental Divide.

Crews diverted the turbulent Chagres River to create a lake at Gatun, near the Caribbean Sea. Finally, in 1909, they began to construct the locks—six in all, three at each end. Each lock measures 110 by 1,000 feet, with concrete walls as much as 50 feet thick and floors up to 20 feet thick. The locks took 4.4 million cubic yards of concrete and four years to build.

As tall as buildings, the locks work as smoothly as sewing machines. At the touch of a switch, 70 holes in the bottom of each chamber drain it or fill it with water from Gatun or Miraflores Lake.

But it is the gates that are the triumph: each 65-foot-wide, 7-foot-thick hollow leaf floats. They close to form a flattened "V" pointed into the force of the water; its pressure keeps them from leaking. The tallest set, at Miraflores on the Pacific side, weighs 745 tons and stands as a monument to precision metalworking.

Culebra (now Gaillard) Cut, a 9-mile-long slice through the Continental Divide, took 7 years and 61 million pounds of dynamite to carve because of repeated landslides.

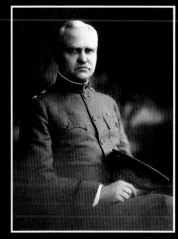

In 1907, Col. George Washington Goethals took over as director of the canal project—after the resignation of John F. Stevens. A stern taskmaster, Goethals once barked when a landslide had buried months of work at the Culebra Cut, "Hell, dig it again." Yet he dealt fairly with complaints from anyone on Sundays from 7:30 a.m. until noon. So his men kept digging, and finished the canal in just seven years.

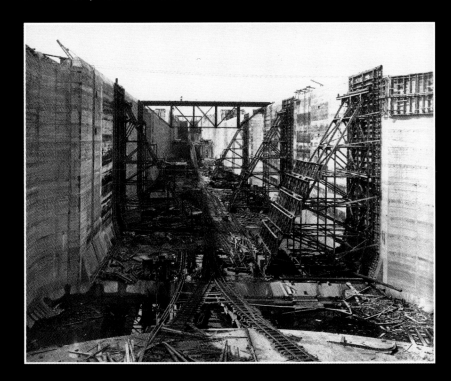

Canals

Water has often provided the most economical and efficient way to transport bulky, nonperishable goods. As Europe's commerce expanded during the Roman Empire, canals appeared. One of the oldest still in use in Britain, Fossdyke, connects the Trent and Witham Rivers at Lincoln.

Rome's collapse in the fifth century A.D. brought a halt to most canal building; the decline continued until the revival of large-scale commerce in the 12th and 13th centuries. But it wasn't until the 18th century and the industrial revolution that canal building truly came into its own. Locks, inclined planes, and water lifts grew increasingly sophisticated, enabling boats and barges to travel over uneven terrain.

When rail hauling began to put national and regional canal networks out of business in the mid-19th century, builders turned to more adventuresome projects— maritime waterways that would chop thousands of miles off global shipping routes.

The French canal promoter Ferdinand de Lesseps spearheaded the work. He built the Suez Canal between 1859 and 1869, began the Panama in 1882, and was briefly linked to Greece's Corinth Canal. All his projects were designed without locks for sea-level travel.

What made these projects desirable was the rapid growth of steamship trade; what made them possible were new machines and techniques developed in the

At the 1869 opening of Egypt's Suez Canal (opposite, top), European heads of state, crown princes, and celebrities, hosted by dignitaries from the Ottoman Empire, sail the passage. On the banks, fellahin watch the celebration that French Empress Eugénie later called her "last happy memory." A modern tanker (opposite) takes the same route on the canal that still plays a role in world politics.

Dramatic gash through solid rock, Greece's Corinth Canal, opened in 1893, links the Aegean and Adriatic Seas, shortening ship routes by about 150 miles. But engineers cut it too thin; today, only smaller ships can squeeze through.

Canals

Threading through the Thousand Islands between the United States and Canada, a "saltie" (below), or oceangoing ship, sails up the St. Lawrence Seaway. Once blocked by rapids near Montreal, ships can now travel some 2,000 miles into the heart of the continent with the help of locks, like the one at St. Lambert (below, right). Mid-continent ports of call include Toronto, Duluth, Detroit, and Chicago.

industrial age that canals had helped foster. Sixty vessels, many of them new steam-powered dredges, dug 97 million cubic yards of earth from the Suez channel. The 10,000-horsepower dredges could move 6 million cubic feet of material a month. Some could even cut rock underwater.

The Corinth Canal, first laid out by Roman Emperor Nero, was cut an average of 190 feet through rock; it, too, could not have been built without new boring tech-

niques and dynamite. But its narrowness makes it navigable only by smaller ships.

Still some of the most used shipping lanes, the Panama and Suez Canals also have shortcomings today. Large aircraft carriers and fully loaded supertankers cannot fit into the Panama's huge locks, and oil tankers cannot pass through the Suez.

An engineering marvel of our time, the St. Lawrence Seaway opened in 1959, linking the Atlantic Ocean and the Great

Lakes. The most concentrated construction took place in one 190-mile stretch around the rapids that blocked the St. Lawrence River, an enormous undertaking. Entire towns had to be relocated. Some 22,000 workers dug 210 million cubic yards of material and poured 6 million cubic yards of concrete to build locks that empty or fill in minutes. Virtually a new North American seacoast, the seaway has become a conduit for the new global economy.

Water-filled containers—like giant bathtubs set on rails—haul canal boats along an inclined plane at Ronquières in Belgium. The mile-long "canal on wheels," inaugurated in 1968, enables barges on the Charleroi-Brussels Canal to negotiate a 220-foot rise—eliminating an eight-mile, ten-hour detour that required passing through 28 locks. Controlled electronically, each 300-by-40-foot container can accommodate several boats or barges as it winches up or down the five-degree slope between canal levels at a speed of about four feet a second. At either end of the ride, gates are raised, and the boats proceed along the canal. Each container weighs more than 5,000 tons and rolls on 236 steel wheels. The carriers are insulated to keep the water from freezing in winter. Built at a cost of some $17 million, the Ronquières Inclined Plane takes 22 minutes to complete a run and each year attracts some 250,000 visitors who watch its progress from a viewing tower.

Aqueducts

G rowing concentrations of people in ancient cities produced the first urban crisis: obtaining an adequate municipal water supply. Needs rapidly outstripped the resources of local springs and wells, and systems had to be devised to bring in water from a distance. Up until the last century, little was known in the hydraulics of water-supply technology that hadn't been tried by the engineers of Babylon, Assyria, Palestine, Asia Minor, or Greece.

By 1000 B.C., water for Jerusalem was being collected from springs in cisterns outside the city and distributed by canals. Mycenaeans and Athenians tapped faraway sources with underground conduits built of stone or cut through rock; they brought water inside city walls. At Pergamon, an aqueduct—perhaps an earthenware or wooden pipe—crossed two deep valleys with sufficient head of water to reach the hillside city; both crossings made use of inverted siphons to push water uphill to its final destination.

But civilized living for a million ancient Romans called for more water—for baths, fountains, and gardens—than had ever been required before. Their engineers developed a true municipal water-supply system, so well organized that modern cities still look to it as a model. The very word "aqueduct" comes from Latin.

Arcades, bridges, siphons, conduits, tunnels, reservoirs, distribution pipes, and water meters formed a network that by A.D. 97 was 880 miles long. It supplied 20 to 400 million gallons of water a day to the imperial city. The exact amount is impossible to calculate because the water commissioners based their figures only on pipe diameters, not the rate of flow. They attributed discrepancies to water theft—in

Ancient Romans went to great lengths to ensure a steady water supply, and their system of aqueducts remains a major engineering achievement. Dramatic, high aqueduct bridges such as the 19 B.C. Pont du Gard in France (opposite, bottom) connected miles of conduits carrying water. Many modern aqueducts, such as one from the 19th century near Malaga, Spain (opposite, top), follow Roman designs. Water still flows through the channel of Segovia's ancient aqueduct (below).

Howling and squeaking as it revolves, a Roman-style *noria*, or waterwheel, in Syria (above, left) uses buckets attached to its rim to lift water. Ancient devices like this, often introduced by the Romans, still supply water for drinking and irrigation in many developing countries. The devices use neither fuel nor expensive parts that need repair.

FOLLOWING PAGES: Majestic arches of Segovia's Roman aqueduct still rise into the Spanish sky. Iron pegs fitted into holes in the granite blocks made transporting them easier. All through their empire, Roman engineers built aqueducts, equating civilization with water.

actuality a considerable problem.

Flowing down from the Appennine hills in an open channel covered by a peaked roof, the water depended on gravity; a downgrade of two or three feet per mile was maintained. When the conduits, which followed ridges most of the way, reached the plain surrounding Rome, they crossed on the long, high bridges we call aqueducts.

As aqueduct was grafted onto aqueduct, or as the weight of water in new channels was added to existing arcades, the structures needed constant repair. After 33 B.C., concrete lined or formed the channels, and was also used to build dams to hold water in reservoirs. Because the engineers were unaware of the principle of thermal expansion and contraction, their water systems leaked. Most of the conduits

lay below ground, where they could be repaired more easily.

When the water reached the city, it flowed into holding tanks to let mud and debris settle out. Then it was piped to a tower, from which it flowed into smaller tanks. Distribution pipes then fed it to public fountains and baths and to private users.

Pipes were made of interlocking, slightly conical sections of terra-cotta or of lead. Lead pipes were made by soldering lead sheets shaped on a core. One official noted that lead pipes weren't healthful, but his warning went unheeded.

Even so, every citizen had access to water, whether collected from his local fountain or piped directly to his villa (by special permit). Not until the 19th century in Europe and America were cities supplied with water on a comparable scale.

Water-Supply Systems

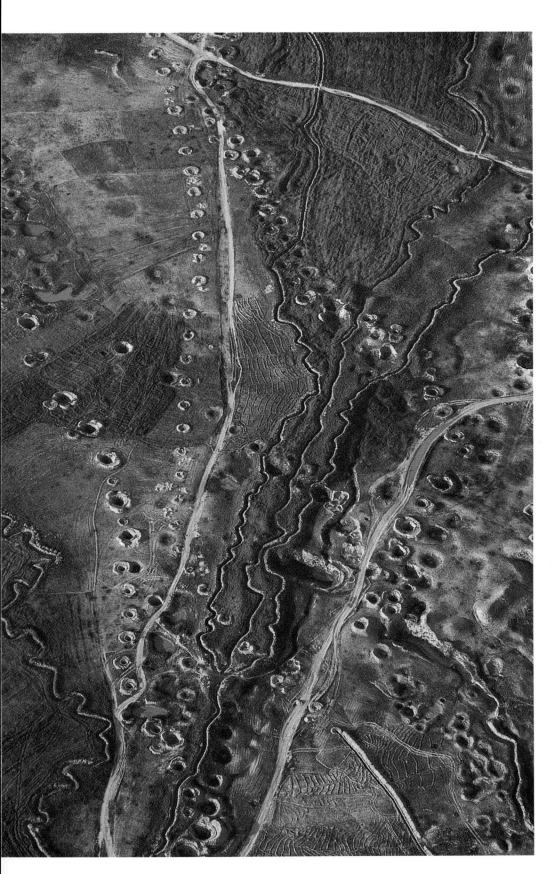

ydraulic engineering got its start as soon as people began to farm the dry deserts of the ancient Middle East. In Mesopotamia, a vast system of irrigation channels brought water from the Tigris and Euphrates Rivers to the fields; building and maintaining the system marked one of the earliest ventures into civil engineering. Large catch basins in Egypt stored floodwaters from the Nile and distributed them by a system of dikes and canals. Babylonian engineers waterproofed the Hanging Gardens with bitumen, the gummy seepage from underground oil deposits.

For more than 2,000 years, in many areas of the Middle East, an ingenious underground system of *qanats* has used gravity to transport water for irrigation. A qanat consists of a sloping tunnel driven

Pockmarks in the desert (left) mark the paths of qanats leading to Firūzābād, Iran. The manholes allow access for maintenance of the sloping underground water channel by means of a vertical shaft (above). Drilled down to the water table in faraway hills, most qanats carry an average of 100,000 gallons of water a day to Iranian towns and fields; elaborate networks can carry up to three million gallons a day, safe from the hot desert sun. As recently as 1933, all of Tehrān's water came from qanats.

into a hillside aquifer or water table. Vertical shafts spaced a few hundred feet apart provide outlets for tunneling spoil, as well as access and ventilation. Under this system, water could flow for miles with little or no evaporation.

A wide range of ancient devices to lift water still serves the region's small farmers. Egyptians use the *shaduf*, a counterweighted lever that lifts buckets — seen in murals from 2500 B.C.—and the Archimedes' screw, invented around 250 B.C. Both are about as efficient as a small diesel-driven pump. The *noria*, a large waterwheel studded with pots, lifts water from stream to aqueduct. The *saqiya*, a refinement developed by Muslim mechanical engineers in the Middle Ages, is a more elaborate geared version driven by oxen or a waterwheel that lifts a steadily revolving chain of pots. The aqueducts carried the water to towns or to garden fountains.

Under the gaze of 12th-century Muslim sidewalk superintendents (left), a *saqiya* lifts pots of water from river to aqueduct. Here, a waterwheel—not the whimsical wooden ox—drives a shaft; gears at the top turn another wheel that operates the conveyor chain of pots.

An Egyptian carpenter fits flanges of an Archimedes' screw (above), an ancient water-lifting device widely used by the Romans. A farmer (below) draws a steady stream of water into his bean field by turning an Archimedes' screw mounted inside a metal pipe.

Bridges

High-rise cables crest a tower on New York's Verrazano Narrows Bridge, one of the world's longest suspension spans.

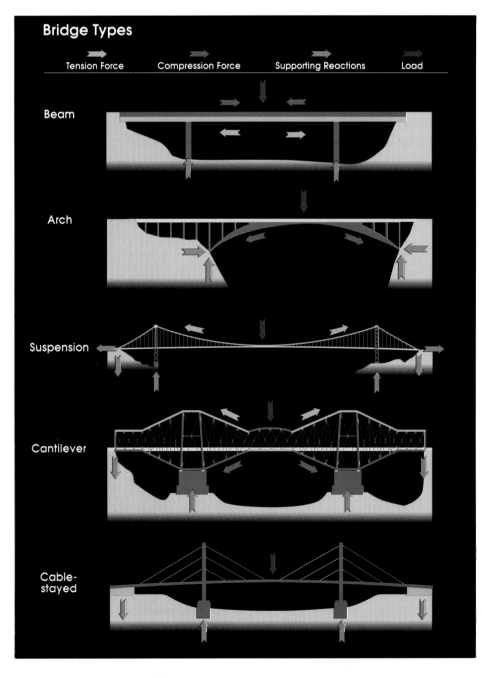

Bridge Types

Tension Force   Compression Force   Supporting Reactions   Load

Beam

Arch

Suspension

Cantilever

Cable-stayed

The first challenge to bridge builders is geography: a valley, river, bay, or other barrier. They must consider not only the length of span necessary to overcome it but also the winds, temperature, and traffic.

After a study of the site, the choice of bridge type falls generally among five: beam, arch, suspension, cantilever, and cable-stayed (diagram). A bridge can be one or a combination of types. The vertical loads on the bridge—its deadweight and the weight of traffic—as well as other forces are shown by arrows.

The beam, or girder, bridge, is simply a span supported at either end like a plank across a ditch. Strength of material limits the length of a single span, but other spans supported on piers may be added. The arch converts vertical loads into axial compression forces that are carried into the ground. The suspension bridge, relatively light and flexible, can leap distances at a bound; cables anchored at either end bring the roadway loads to towers, which carry them to the foundations.

In the cantilever bridge, two arms extend from opposite supports, each fixed at one end only. They sometimes carry a shorter span in the middle. Engineers introduced such bridges for railway loads. In a cable-stayed bridge, cables connect the roadway directly to a tower, which carries the load to the foundations.

Beam

The modern Quesnell Bridge (above), in Canada's Alberta Province, contrasts with China's An Ping Bridge (below), from the 12th century. While steel and concrete support the one and granite the other, both follow simple beam bridge principles. The An Ping Bridge's piers are shaped to withstand the assault of tides.

Drop a tree across a stream or place a stone slab across a ditch and you have the essence of the beam bridge—a horizontal span supported at either end. Materials may have changed over the years, but not the basic structure itself. And while the origins of such bridges are lost in antiquity, examples of them remain the world over.

In England, stone "clapper" bridges, some perhaps dating back to ancient Celtic times, cross streams on the moors of Devon. In China, the 800-year-old An Ping Bridge carries villagers nearly a mile across an estuary along the southern coast. To place the granite beams, some 15 feet long and weighing up to 100 tons, builders had to float them into position during the

long distances with relatively little material. Iron and then steel trusses soon replaced many of the flimsy wooden trestles that had been used to carry rails across rumpled terrain. Trusses are like beams in which thin, vertical posts and diagonals make up the area between the top and bottom horizontal members.

The Britannia Railway Bridge, a giant, rectangular iron tubular beam built across the Menai Strait in Wales in 1850, became the precursor of the popular, modern box-girder bridge, so called because its hollow cross section acts like a beam.

Today, reinforced and prestressed concrete—both developed since the late 1800s—are widely used in beam bridges.

highest tides of spring and fall.

The advent of relatively inexpensive cast iron and of wrought iron (iron heated and hammered into a desired shape) and the rapid spread of railroads in the mid-1800s gave rise to a golden age of bridge building: Heavily laden trains could negotiate only the gentlest inclines, requiring their routes to be as flat as possible. This led to long viaducts, often with wide spans and high supporting towers. Truss bridges became popular because they could span

Granite piers and six-foot-wide slabs form this "clapper" bridge in Devon, England. Such bridges, among the world's oldest, get their name from *claperius*, Latin for "pile of stones."

Lethbridge Viaduct in Alberta, Canada (opposite, upper), a mile long and 314 feet high, replaced an earlier rail line that crossed the valley of Oldman River on 22 wooden bridges.

Chesapeake Bay Bridge-Tunnel (opposite, lower), completed in 1964, carries the elevated part of its two-lane roadway on some 3,000 pilings made of prestressed concrete.

Arch

The Romans regarded bridge building as a sacred calling, to be entrusted to a special priestly class headed by a Pontifex Maximus. They were not the first to erect arch bridges, however; the Mesopotamians had built them some two thousand years before. The Roman arch was semicircular and so carefully crafted that its own weight held it together without mortar. A temporary wooden frame supported the arch until the keystone was added, locking the structure together. But such bridges had relatively narrow archways, an impediment to navigation.

Half a world away, a builder named Li Chun solved the problem around A.D.

610 by flattening out his arch to build the Great Stone Bridge across the Jiao River in northern China. Such flattened arches made longer bridge spans possible.

Medieval Europeans also used stone arches in bridge construction. London Bridge, one of a succession of spans that have occupied the site since Roman times, crossed the River Thames on 19 arches.

The industrial revolution of the 18th

Great Stone Bridge (above) arcs across China's Jiao River. Arched openings act as spillways, reducing water pressure during floods.

West Yorkshire's Wycoller Bridge (above, left), from the early 1800s, resembles a Roman span.

Encrusted with houses and shops, Old London Bridge stood, despite misuse and neglect, for more than 600 years. Begun in 1176 by Peter of Colechurch, a priest and engineer, the bridge took 33 years to build across the 900-foot-wide River Thames, which, reportedly, was diverted four miles during construction. The bridge had a draw, 19 pointed stone arches, and sturdy piers on foundations pointed to resist the tides.

Arch

Versatile dreamer and risk-taker, Isambard Kingdom Brunel (1806-1859) was the son of Marc Brunel, a British engineer of international reputation. Young Brunel applied his genius to building tunnels, bridges, railway stations, ships, docks, and water towers—many still in use. He designed the Royal Albert Bridge at Saltash and London's Paddington Station; he introduced broad-gauge railways and initiated the modern era of transatlantic steamships. One of his ships laid the first transatlantic cable.

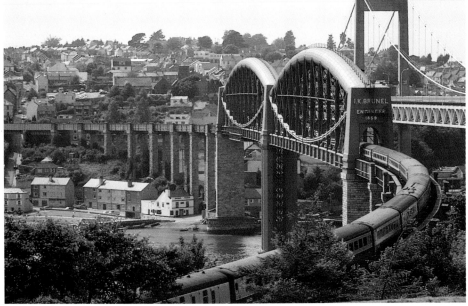

A contemporary engraving and a modern photograph show Saltash Bridge, also known as the Royal Albert Bridge, to have changed little since its May 1859 dedication in Cornwall, England. Daringly combining both arch and suspension spanning features, I. K. Brunel built two main spans and 17 shorter approach spans. Each 1,060-ton main span was fabricated on shore and floated into place on

huge twin pontoons. Hydraulic jacks then lifted the spans 100 feet into place on the piers.

Rare survivor of a century-long era of cast-iron bridge building, Ironbridge (opposite), on England's Severn River, weighs 378 tons and took only three months to assemble in 1779.

and 19th centuries brought iron—either cast or wrought—to bridge building. Ironbridge, erected by Abraham Darby in 1779, spans the Severn River near Coalbrookdale, one of England's earliest iron-making centers. "Though it seems like network wrought in iron," marveled an observer, "it will apparently be uninjured for ages." Indeed, Ironbridge still carries cyclists and pedestrians.

The bridge emulates in cast iron the principles of carpentry. No bolts or rivets originally held Ironbridge together; its joints and fastenings were metal equivalents of the slots, dovetails, and mortise-and-tenon grooves used by woodworkers.

When a flood in 1795 demolished

most of the Severn's wooden bridges, Ironbridge stood fast, enhancing iron's reputation for strength and durability—and leading Thomas Telford to turn to iron. Telford, one of Britain's great civil engineers and a structural artist, built iron bridges—such as the 1814 cast-iron bridge over the River Spey in Scotland—that remain as models of their time.

In France, Gustave Eiffel, better known for his later tower in Paris, was also winning contracts to build iron bridges. His Garabit Viaduct over the windswept gorge of the Truyère River carried a single railway line some 400 feet above the valley floor.

To reduce wind resistance, Eiffel used open-truss girders. Employing a more parabolic shape and building with wrought iron, Eiffel created modern forms that still serve today. He built hinges into the arch's base that allowed flexing when the temperature changed. For many years after its completion in 1884, Eiffel's bridge stood as the world's highest arched span—a distinction now going to the 700-foot-high Glen Canyon Bridge over the Colorado River.

And in 1855, Isambard Kingdom Brunel, another towering British engineer, had begun work on the Saltash—or Royal Albert—Bridge over the Tamar River near

Plymouth, England. A composite span, known as a lenticular truss, it combined elements of three major bridge types—arch, suspension, and (on the approaches) beam. Twin tubular arches of wrought iron, each spanning 455 feet, are connected at their ends to link chains. The arches and chains jointly carry the deck.

Brunel used a compressed air caisson to dig down to solid rock so that the pier foundations could be built. James Eads did the same in 1874, when he constructed a railroad bridge across the Mississippi River at St. Louis, the first large structure built of steel. Its three arch spans

Arch

were, at the time, the longest in the world.

In more recent years, concrete reinforced with steel has led to spectacular arched spans, including Eugène Freyssinet's 1930 bridge at Plougastel, France, and Swiss engineer Robert Maillart's elegant Alpine bridges, such as the 1930 hollow-box bridge over the Salginatobel River, near Schiers, and the 1933 deck-stiffened arch over the Schwandbach River. Maillart

created stiff structures by fusing very thin concrete arches with the horizontal deck.

The 20th century's longest arch bridges are made of steel. Though often lighter, they resemble reinforced concrete bridges in design and function. Notable examples include Othmar Ammann's 1931 Bayonne Bridge, with its 1,652-foot span, and the 1981 New River Gorge Bridge in West Virginia, which spans 1,700 feet.

The open-truss wrought-iron girders of Eiffel's 1884 Garabit Viaduct (left) allow wind to pass.

In Schwandbach Bridge (above), Robert Maillart's 1933 concrete masterpiece in Switzerland, cross walls tie the thin, vertically curved arch to the horizontally curved roadway.

Suspension

Unsurpassed in length of span, the suspension bridge arcs gracefully across empty space to connect distant or hard-to-reach places. Major spans, such as that of Japan's Akashi-Kaikyo, scheduled for completion in the late 1990s, can vault more than a mile.

Four elements distinguish a modern suspension bridge: its roadway, also called the deck, which usually stretches over the main span and the two side spans; towers at both ends of the main span; cables slung over the tops of the towers; and solid anchorages, normally at the ends of the side spans. From the main cables hang vertical supports called suspenders, or hangers, which carry the deck weight and its live load up to the cables. On some bridges, saddles on top of the towers allow the cables to adjust to temperature changes and shifts in load.

Anchorage blocks, constructions of steel and reinforced concrete, hold the cable ends in place. One such block for the 4,625-foot Humber Bridge in England, opened in 1981, measures 213 feet long and 118 feet wide. The weight of these anchorages resists the pull of the cables.

Rising through the mist, the Golden Gate Bridge spans the entrance to San Francisco Bay in one 4,200-foot stride. Completed in 1937, it has the tallest towers of any bridge; trusses beneath the roadway stiffen the deck.

A bridge of braided grass (opposite, left) crosses a gorge in Peru, recalling Inca spans that once laced such ravines. Some 22,000 feet of cordage, made by spinning *coyo* grass stalks, are twisted into ropes, which are then braided into cables and pulled across the gorge.

An Andean worker tightly strings ropes between the hand and foot cables of a grass suspension bridge (opposite, right), to prevent even small pedestrians from falling through. The bridge will need replacing in a year.

The Brooklyn Bridge

B rooklyn Bridge, in its day the world's longest suspension span, crosses New York's East River to link Manhattan and Brooklyn. Designed by John Roebling, a German immigrant engineer and creative genius, the bridge took 14 years to build—at a cost of $15 million and at least 20 lives, including Roebling's own. His son, Washington, finished the job. The bridge deck arcs

Crippled and in great pain from the bends, Col. Washington Roebling sits by his sickroom window in Brooklyn Heights, directing work on the bridge envisioned by his father. Binoculars enabled him to supervise the project. After the elder Roebling's death in 1869—from a tetanus infection—his son, himself an accomplished engineer, took over the project. "Here I was at the age of 32," he later wrote, "suddenly put in charge of the most stupendous engineering project of the age!" When caisson disease disabled him, his wife, Emily, learned engineering and acted as his spokesman. He eventually recovered and lived to the age of 89.

technique recently invented in Europe. Two inverted, airtight timber boxes—each half as big as a city block—were dug down to bedrock and filled with concrete.

Laborers digging at the base of the caissons worked in air chambers pressurized to keep water from flooding in. More than a hundred workers—known as sandhogs—were stricken with the bends because its cause and prevention (by use of decompression chambers) were not then understood. Three of them died.

The bridge's four main suspension cables used steel wire—nearly 15,000 miles of it—instead of the traditional wrought iron. To thread the wire back and forth across the tower tops, John Roebling invented a "traveling-wheel" rig, used by bridge engineers ever since. Sailors accustomed to working in the high rigging of sailing vessels wrapped the cables.

Each completed cable measured nearly 16 inches in diameter. It contained 19 wire bundles called strands. Each

The world's first electrically lighted bridge opened May 24, 1883, amid great fanfare. Its 85-foot-wide deck originally carried carriageways, cable railway tracks, and an elevated promenade (below). Today, complex interchanges funnel six traffic lanes across the bridge; a central walkway remains, but the tracks came out in 1944.

1,595 feet between towers that rise 276 feet above the river. Side spans, each 930 feet long, reach from tower to shore.

Hailed as the eighth wonder of the world when it was completed in 1883, the Brooklyn Bridge required prodigious feats of engineering. To sink piers for the massive granite towers, Washington Roebling used the pneumatic caisson method, a

The Brooklyn Bridge

strand, in turn, consisted of 278 individual wires. By laying looped cable wire over the towers one loop at a time, the Roeblings avoided hoisting heavy cable, as well as the risk of damaging the tower. Another of John Roebling's innovations—slanted cables, or stays, leading from the tower to various points along the bridge deck—helped hold the roadway steady in heavy

A snapping cable claims two lives in this contemporary lithograph.

A traveling wheel threads wire cable as it shuttles back and forth across the river.

winds. With a clearance of 135 feet between the water and the road, all but the tallest sailing ships could pass freely beneath the bridge.

The bridge united two rival and independent cities, in effect helping to establish the greater metropolis of New York City. More than 100,000 cars a day cross the bridge now, far more traffic than the designers ever envisioned. Yet, so carefully was it conceived that only a few light trusses had to be added to accommodate modern traffic conditions.

To this day, New Yorkers stroll, bike, or jog along the elevated walkway (restored in 1982) that the elder Roebling originally created to "allow people . . . to enjoy the beautiful views and pure air."

Gothic towers webbed with steel (opposite) soar 276 feet above the East River.

Excavated from below and weighted on top by granite blocks, a timber caisson inches downward toward bedrock.

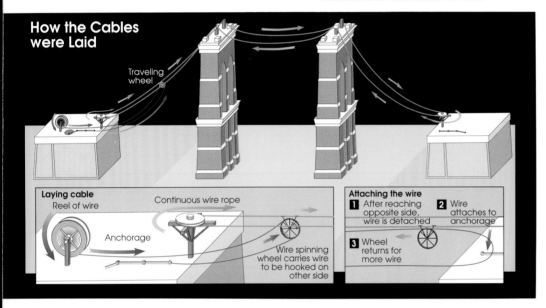

How the Cables were Laid

Traveling wheel

Laying cable
Reel of wire
Continuous wire rope
Anchorage
Wire spinning wheel carries wire to be hooked on other side

Attaching the wire
1 After reaching opposite side, wire is detached
2 Wire attaches to anchorage
3 Wheel returns for more wire

FOUNDATION LINE

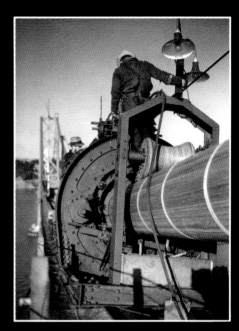

Photos from 1936 show the building of the bridge. Here, a radial jack compresses wire strands into a single cable. A wrapping device binds the cable at three-foot intervals.

Workmen (right) loop steel wire onto the wheel that will string it across the towers.

Workers check the flanged wheels that shuttle loops of wire from one bridge end to the other, using the method devised for the Brooklyn Bridge. Each wheel makes thousands of trips.

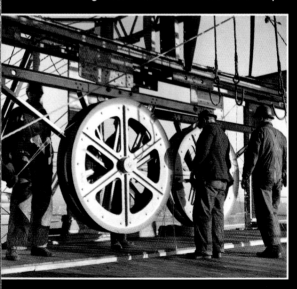

The War Department was adamant: The proposed bridge across the mouth of San Francisco Bay could not hinder navigation. Nor could it interfere in any way with area military facilities, including a nearby Navy yard. And it had to be earthquake proof.

Chief engineer Joseph Strauss settled on the most likely choice: a long, thin, flexible suspension span that would, at 4,200 feet, leap almost all the way across an expanse far greater than any previousl attempted. And although Strauss had built nearly 400 bridges around the world, he had never tackled a suspension bridge.

The War Department approved the project in December 1924, but numerous

Gate

lawsuits—and the 1929 crash—delayed groundbreaking until January 1933.

The north tower, built on a rock ledge close to land, went up with little trouble. But foundations for the south tower, in deep water 1,125 feet from shore, had to be built inside a giant cofferdam workmen dubbed "the bathtub."

Here, on the threshold of the open ocean, powerful tides swept in and out twice a day at better than seven knots. It was only during the turn of the tides—four daily periods of 20 minutes each—that the water was still enough for underwater work.

Storms and a ramming by a fog-bound freighter were among the hazards encountered—and overcome. Ten men died when their scaffold collapsed, hurling them into the sea 240 feet below.

Four years after it was started, on May 27, 1937, the bridge was opened to traffic. "At last the task is done," wrote Strauss. But the task had taken its toll on him. Within a year, at age 68, he was dead.

The north tower of steel rises 746 feet.

Eyebars embedded in gigantic concrete blocks hold the splayed ends of a bridge's cable, as here on the Oakland Bay Bridge.

Suspension

James Finlay built the first American suspension bridge, in 1801. He guaranteed his 70-foot-long span, erected in Pennsylvania, "for fifty years [all but the flooring]." Two decades later, England's first vehicular suspension bridge—Union Bridge—crossed the Tweed River at Berwick, and in 1826 Thomas Telford built the first modern suspension bridge across the Menai Strait in Wales, with a main span of 580 feet. Thousands of wrought-iron links composed its cables and suspenders.

The most influential 20th-century suspension span was Othmar Ammann's George Washington Bridge, across the Hudson River between Manhattan and New Jersey. Completed in 1931, the bridge was extraordinary in weight and proportions: Its 3,500-foot main span doubled the length of the earlier record holder. Most unusual was the anchorage on the New Jersey side, where a rocky cliff rises some 225 feet. One hundred and sixty feet below its top, workers blasted two tunnels through which the bridge cables were passed and anchored in the rock. Ammann designed the bridge's heavily trussed steel towers to be covered in concrete and granite, but the Depression intervened, leaving the towers bare.

The heavens illuminate Michigan's Mackinac Bridge. Completed in 1957, its arching roadway vaults 3,800 feet between towers.

Marathon runners crowd the 1964 Verrazano Narrows Bridge across New York Harbor (right, upper). Designed by O. H. Ammann, one of the century's greatest bridge builders, the double-decked Verrazano carries 12 traffic lanes.

The 5.8-mile Great Seto Bridge (right, lower), opened in 1988, crosses Japan's Inland Sea to connect the islands of Honshu and Shikoku.

Cantilever

Ill-fated Quebec Bridge collapsed twice during construction, killing 86 workers. Designed to set a record for steel cantilever spans at 1,800 feet, the railway bridge was nearly finished in 1907 when 20,000 tons of steelwork crashed into the St. Lawrence River. An investigation showed numerous design deficiencies. Here, a new central span falls into the water while being hoisted into place in 1916. The bridge finally opened to traffic in August 1918.

In a cantilever bridge, two beams project from opposite piers. Each beam is supported by its pier and anchored at one end, like a swimming pool springboard. In such a bridge, the projecting beams may meet in the middle, forming a rigid, continuous beam. Some cantilevers have projecting spans that support a shorter, central span. Such is the case with Scotland's Forth Bridge. Cantilever bridges are generally stiffer than suspension bridges and hence suitable for heavy traffic such as a railroad train with its rhythmic, pounding thrust.

Designed by Benjamin Baker and completed in 1890, the massive mile-long Forth Bridge near Edinburgh was an engineering triumph of its time—one of the first major railroad bridges to use steel and, until the opening of the Quebec Bridge in 1918, the world's longest span.

On three massive piers, cantilevers form two main spans, each of them 1,710 feet long, about 150 feet above the water. The steelwork includes more than 50,000 tons of tubes and girders.

But such bridges are expensive to build and maintain. The dawn of the automobile age and the development of reinforced concrete in the late 19th century ushered in a new era of building all types of bridges. Here was an inexpensive, strong material that required little upkeep.

In 1928, Eugène Freyssinet developed another method for uniting steel and concrete. In prestressed concrete, steel tendons are fitted into a concrete beam, stretched to a high-tension stress, and anchored at either end. The tendons pull the ends together, pushing the beam inward and upward. The upward force helps counterbalance the downward force of bridge loads, while the inward force

counteracts the tension due to gravity loads. Prestressed concrete is often easier to use and requires less material than reinforced concrete; in the 1940s, it became a major building medium.

The cantilever method of erecting prestressed concrete bridges, developed by German engineer Ulrich Finsterwalder in the early 1950s, consists of building a concrete cantilever in segments, prestressing each one onto the earlier ones, and supporting the next one from the previous segment. This avoids the need for scaffolding and

Tubes, some big enough for a train to roll through, strengthen the massive towers of Scotland's Forth Bridge, a masterpiece of steel design. Three huge caissons embedded in the river and filled with concrete support the towers. Designer Benjamin Baker favored cantilevers for spans of more than 700 feet.

Cantilever

Ganter Bridge (above and opposite), opened in 1980, traverses a Swiss Alpine valley. The bridge combines both cantilevers and cable stays for strength and grace. In melding technique and aesthetics, Christian Menn's design followed a bridge-building tradition that began nearly two centuries ago with iron arches.

was used, for example, on Germany's 1964 Bendorf Bridge and the Ganter Bridge near the Simplon Pass in Switzerland.

With the Ganter, Swiss engineer Christian Menn introduced in 1980 a new, hybrid bridge form. Cantilever spans supporting the bridge's midsections are themselves supported by cable stays embedded in prestressed concrete. These concrete walls add safety through their strength and protect the cables from stress and corrosion. Menn's main motive, however, was aesthetic: Like a sculptured bird, the Ganter Bridge appears to take wing.

In Guangxi, China, successively longer timber cantilevers support a roofed bridge.

Cable-Stayed

Its towers connected directly to the deck by steel cables, the cable-stayed bridge represents a modification of a technique long used to strengthen suspension bridges. John Roebling included cable stays in his design for the Brooklyn Bridge, along with vertical suspenders.

Cable-stayed bridges have reached spans of 1,700 feet, and soon they will nearly double that. They have proved popular for medium-length bridges because they do not require the large anchorages suspension bridges need, and because they can be economically built by the cantilever method. The diagonal steel cables called stays attach the decks—often box-girder structures of prestressed concrete—to the tall towers or masts.

German engineers pioneered the design of cable-stayed bridges in the 1950s and 1960s, especially for Rhine River crossings destroyed during World War II. Particularly impressive is a series of light, harp-type bridges (with parallel cables) built as a family at the bend in the Rhine River at Düsseldorf.

In the United States, a noteworthy cable-stayed bridge was completed in 1987 across Florida's Tampa Bay. The 4.1-mile Sunshine Skyway Bridge carries four lanes of traffic on a divided road that sweeps 175 feet above the main channel.

The bridge's precast, prestressed roadway sits atop massive twin piers 175 feet high, 8 other main span piers, and 36 approach piers. Two pylon towers, built over the main piers, support the 42 cable stays and direct the weight of the main span to the foundations. Around the central part of the bridge, protective bumpers known as "dolphins" can absorb the force of an 87,000-ton ship moving at 10 knots.

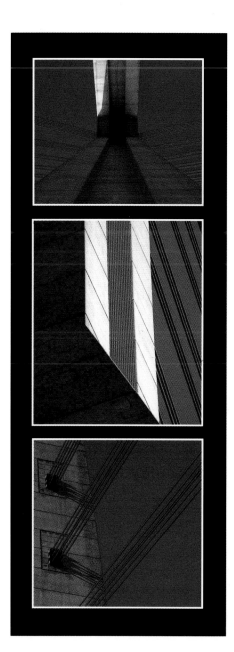

Steel cables support the 1,200-foot main span of the Sunshine Skyway Bridge (left), in Tampa Bay's major shipping channel. Designed to withstand hurricane-force winds, the bridge has many safety features, including barriers to protect piers from off-course ships and an electronic warning system for motorists. Near Flehe, Germany, cable stays (above) grace a 1979 Rhine River bridge.

Railroads

A 19-mile rock causeway carries the Southern Pacific Railroad across Utah's Great Salt Lake.

CELEBRATING THE COMPLETION OF THE TRANSCONTINENTAL RAILROAD, 1869.

On the Trans-Australian Railway, trains crossing the treeless Nullarbor Plain shoot across the world's longest stretch of straight track, 300 miles without a single curve. But few locations on earth offer the long, flat, straight stretches that permit locomotives to pull heavy loads at top speeds. The challenge to the railroad engineer has been, from the first rail lines in the 1820s and 1830s, to reduce the factors that slow trains down—grades and curves—by flattening and straightening the roadbed. Railroads work best with grades no steeper than 2 percent; tunnels, bridges, and landfill all help keep grades to a minimum.

The railroad engineer also seeks to avoid grade crossings—intersections with highways or other railroad tracks—which force a train to slow down. And engineers the world over have had to find ways to go around, over, or through forests, swamps, mountains, rivers, and deep valleys.

Locating the line is crucial. A well-chosen route might bring the railroad great profits; a badly planned road could easily lead to financial ruin. Land that looks easy can still present serious problems. The locating engineers—route surveyors sent out to examine the terrain and choose a route—would balance difficulty of construction (and thus cost), safety, efficiency, and operating expenses as they pounded in the stakes that marked the line.

erhaps no other engineering feat has thrilled Americans as much as the building of the 2,000-mile transcontinental railroad. Surely no other project overcame such physical and logistical nightmares: Rails for the Central Pacific, building east from Sacramento, had to be shipped around Cape Horn. The Union Pacific Railroad, starting west from Omaha, Nebraska, had to bring in 2,400 wooden crossties for each mile of track because there were few trees on the plains. Attacks by Indians meant every rail crew had to keep carbines handy.

The Central Pacific faced labor shortages. When the first 50 Chinese workers proved to be hard-working and tenacious—as well as cheap labor—thousands more were hired. Crossing the Sierra Nevada, it was the Chinese who blasted out the ledges along the steep gorges. Lowered in baskets, they drilled holes in the rock, tamped powder, set the fuses, and yelled to be hauled up quickly.

Tunneling through the hard granite of the high Sierras was complicated by the harsh winter of 1866-67. Blizzards piled drifts so deep that 12 locomotives could not

the Transcontinental

push a snowplow through. Miles of snow-sheds had to be built to protect the tracks.

Tracklaying went faster in the desert, despite heat and water shortages. Rewarded with federal loans and land grants for every mile completed, the two companies passed each other near Ogden, Utah, and kept right on building. By the time the government ordered them to join up at Promontory, 225 miles of parallel track had been laid.

On May 10, 1869, as the last spikes—of gold—were driven and locomotives from east and west touched cowcatchers, telegraphers flashed the news across the country: "Done."

Union Pacific workers (opposite), on the largely treeless Great Plains, dress timbers shipped from the East at great cost by rail and boat.

Cutting corners to speed tracklaying, workers often built quick, temporary trestles (left). A masonry bridge later replaced this one at Green River, Wyoming.

Layer by layer, grading crews blasted and hacked away huge volumes of rock to achieve a minimal grade (above). Here a crew nears Ogden, Utah. Tracklayers were close behind.

Railroads

Two engines pull a Denver and Rio Grande train up Colorado's San Juan Mountains in a photograph from the 1880s. A decade and a half earlier, Swedish chemist Alfred Nobel had invented dynamite, making ledge blasting easier and safer than with earlier explosives.

In 1850, railroads were under construction all over the eastern United States; they expanded westward in the 1860s. By 1870, most of Europe's main lines had been built, and by the late 1800s, networks existed in much of the rest of the world.

When competing with other railroad companies, locating parties often moved with secrecy and speed. In the 1880s, Edward Gillette of the Great Northern reportedly perfected the three-day method. His party would locate the line the first day, purchase the right-of-way the second day, and on the third day "there would heave

barrows, and one-horse dump carts. Often geological features that weren't in the way were blasted to create fill for others that were. Blasting powder usually did the job. Dynamite did not come into use until 1867. Nitroglycerine, a powerful but notoriously unstable liquid explosive, was sometimes used. Carelessly handled, a half-cup could blow a work gang to smithereens.

Right behind the grading crew came workers known as gandy dancers for the Chicago company whose tools they used; they laid wooden ties along the grade and then the rails. In England and the United States, the first rails had been laid on granite blocks. In 1830, American engineer Robert Stevens laid rails on a bed of heavy wooden crossties, or "sleepers," and discovered that not only was the wood cheaper, it also gave a smoother ride. Spiked to the ties, the rails were connected to each other with steel pieces called fishplates. Then the line would be ballasted— the spaces between and under the ties filled with crushed rock so that the track would not shift under the trains. Gradually, iron rails were replaced by steel ones, which last 20 times longer.

When topography couldn't be rearranged, it was finessed. Railroad builders in India in 1859 faced the steep Western Ghats south of Bombay. The absence of ledges and plateaus made the mountains seem impassable. To scale them, engineer James Berkley routed the rails up the mountains in a series of switchbacks, or tight zigzags, with a reversing station in each angle. Eventually a tunnel replaced this tortuous mountain climbing.

When the Norfolk and Petersburg wanted to cross the Great Dismal Swamp in the 1850s, it sent in 26-year-old William

Inching along the same century-old road to riches, the Durango and Silverton Railroad now carries tourists, not miners, through the steep San Juans. Only three feet wide, narrow-gauge lines suited Colorado's mountains, where trains often wound through canyons "so crooked a bird couldn't fly through."

in, as if dumped down from the sky, a crew of several hundred laborers."

As soon as a route was marked, grading began. Crews cleared the right-of-way, usually about 60 feet wide, cutting trees, digging out stumps, and shifting earth with picks and shovels, wheel-

Railroads

Mahone, later a Confederate general. Mahone and his men hacked a 100-foot-wide clearing through the tangled vegetation, then cut deep drainage ditches on either side of the marsh and lined them with granite. They then layered crisscrossed tree trunks over the clearing and piled on earth to create the railbed.

In southern California in 1876, William Hood designed an engineering marvel known as the Great Tehachapi Loop. It brought the Southern Pacific Railroad from the floor of San Joaquin Valley up

and over the Tehachapi Mountains with a series of gradual loops and 18 tunnels, in which the track actually snaked back over itself. The sharp rise of 2,734 feet from valley floor to summit, a distance of 16 miles as the crow flies, was spread out over 28 miles of track with an average grade of only 2.2 percent.

Over the past 50 years, railroad technology has changed rapidly. Engineers have developed sophisticated mobile equipment to mechanize all phases of tracklaying and repair and to locate prob-

lems on the lines. In 1964, Japan opened a new age of high-speed rail travel with the New Tokaido Line, which runs streamlined "bullet trains" between Tokyo and Osaka at an average speed of over 100 miles per hour. A specially designed roadbed—flat, straight, raised above all grade crossings—carries the tracks through an urban area that contains 40 percent of Japan's people and 70 percent of its industry. Modern railroad engineering serves notice: Ideal conditions—flat, straight, no crossings—can be created anywhere.

Flat and straight, a specially engineered roadbed allows "bullet trains" to travel 130 miles per hour on Japan's New Tokaido Line. Tracks run free of hindrances on an embankment high above other traffic arteries.

A late-model tracklaying machine (opposite) upgrades a Burlington Northern line in Wyoming with ribbon rail laid on concrete ties. With fewer joints, quarter-mile rail lengths give a smoother ride and fewer "clickety-clacks."

Pipelines

The Trans-Arabian Pipeline, completed in 1950, stretches 1,068 miles between the Persian Gulf and the Mediterranean Sea.

Workers dig trenches and prepare the "Big Inch" for burial as part of a World War II project to assure northern U.S. factories adequate fuel. Measuring 24 inches in diameter, "Big Inch" carried crude petroleum 1,341 miles between Longview, Texas, and central Pennsylvania. Extensions fed refineries in Philadelphia, New Jersey, and other industrial centers. A smaller pipeline, 20-inch "Little Big Inch," ran parallel and carried refined petroleum products such as heating oil and gasoline. Between them, the two pipelines carried half a million barrels a day.

ipelines are like arteries: seldom seen but crucial to life in a modern industrial world—or even a not-so-modern world. Underground conduits have irrigated parts of the Middle East for thousands of years, and a gravity-fed water system supplied ancient Rome. As long ago as 1580, primitive pumps distributed piped water in London. By the mid-1800s water lines came into widespread use in the United States.

America's first oil pipeline was only two inches in diameter. Built in 1865 by Samuel Van Syckel, it carried the equivalent of 800 barrels a day from a Titusville, Pennsylvania, field to a railroad five miles away—and at far less cost than in barrels.

Since then, engineers have worked to build pipelines of ever greater diameter and length. Steel pipes, and the introduction of electric arc welding in the 1920s, helped usher in modern pipelaying. Advances in pump technology made it possible to move oil cheaply and quickly at pressures of up to 2,000 pounds per square inch. Today, a mostly subterranean pipeline network extends some 2.5 million miles across North America, carrying natural gas or petroleum from supplier to consumer.

"Big Inch" (left), the nation's first large-diameter, cross-country oil line, came into service in 1943. Its chief engineer, Burt E. Hull, also helped build the Trans-Arabian Pipeline (opposite).

Sunlight radiates from an elevated section of the Trans-Alaska Pipeline, whose high-tech design is unsurpassed. Made in Japan, the 48-inch pipeline came in 40-foot lengths and can carry 2.14 million barrels of oil a day. At the height of construction in the summer of 1975, some 20,000 men and women worked 12-hour days, 7 days a week, in a race against winter.

The Trans Alaska

Welders work on a feeder line near Prudhoe Bay (upper and lower).

Cable restrains a length of pipe (right) as it pitches toward the Sikanni River.

Spanning Alaska's girth, the Trans-Alaska Pipeline traces an 800-mile route between Prudhoe Bay and a huge oil terminal in Valdez.

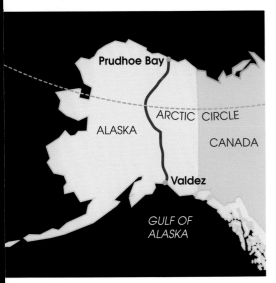

It was, in oilman's parlance, an elephant field. Or, "one of the largest petroleum accumulations known to the world today," as geologists put it, when the magnitude of the Prudhoe Bay find became known in June 1968. But tapping the nearly 10-billion-barrel deposit would not be easy.

Prudhoe Bay, where Alaska's North Slope meets the frigid waters of the Arctic Ocean, is also one of the world's remotest, most hostile environments. Here, far above the Arctic Circle, winter winds howl and temperatures often hover around minus 60°F. And in the brief summers, the permafrost soils thaw just enough to become tractor-swallowing quagmires and breeding grounds for some of the world's bloodthirstiest insects.

Add to that the fact that the pipeline had to scale three mountain ranges, cross more than 350 rivers and streams, and traverse zones of intense seismic activity—

The Trans Alaska

Liquid ammonia circulating to aluminum radiators atop vertical supports (diagram and inset, below) dissipates heat and helps keep permafrost frozen and rigid. Teflon-coated "shoes" strapped to the pipeline enable it to expand and contract freely on the cross-beams. Some 78,000 supports elevate the line for more than half of the 800 miles.

Gate valves as big as men, operated remotely from Valdez, were designed to function at temperatures as low as minus 70°F and can shut down oil flow within four minutes.

Insulated with fiberglass, pipeline buried 3 to 12 feet deep enables caribou to migrate unhindered. In areas of unstable permafrost, refrigerated brine keeps the soil frozen solid, so the pipeline won't sag under its own weight. In streams and rivers, the pipeline is weighted down with concrete blocks (opposite, lower).

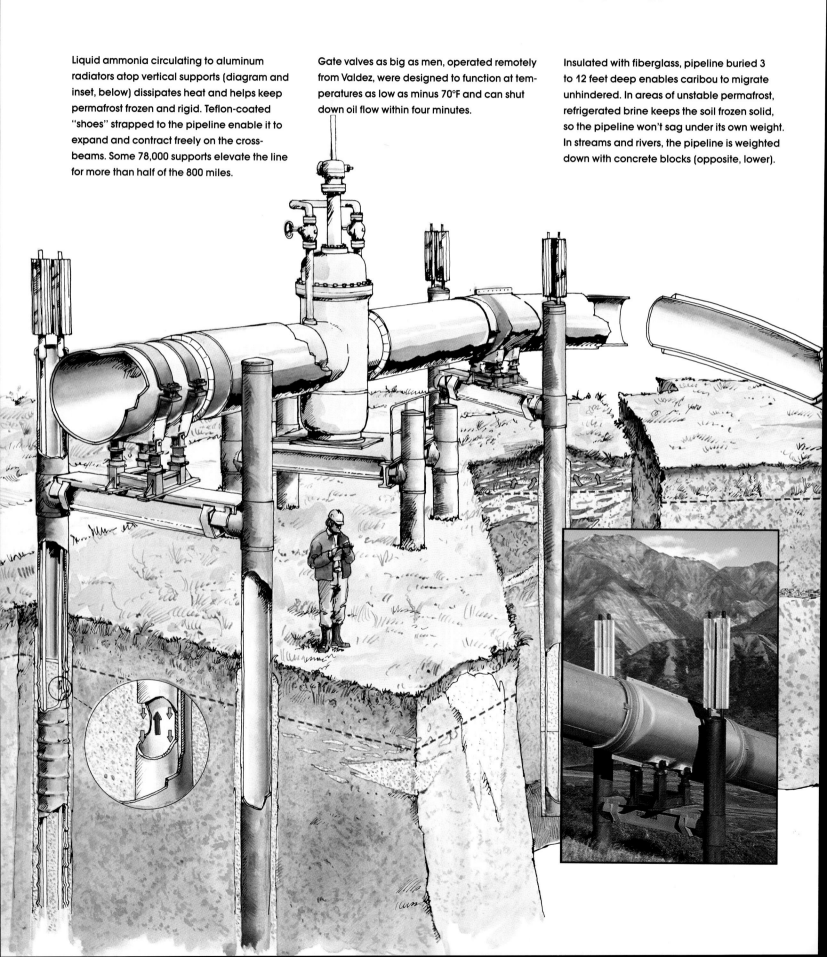

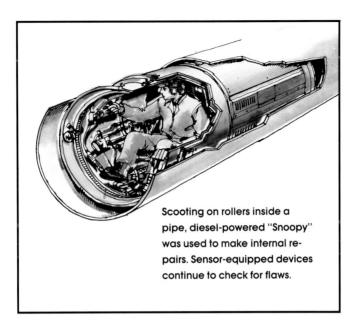

Scooting on rollers inside a pipe, diesel-powered "Snoopy" was used to make internal repairs. Sensor-equipped devices continue to check for flaws.

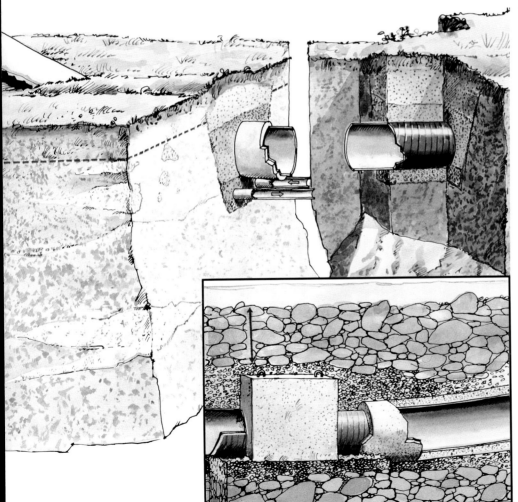

as well as overcome environmental and political concerns—and the task became gargantuan. To build a pipeline under such conditions required truly innovative solutions. Even to get started on the three-year, eight-billion-dollar project took an oil embargo and a 1973 act of Congress. To carry out the task, eight large, multinational oil companies formed an operating group known as Alyeska.

For more than half the route, the pipeline is elevated on what became known as the "high-rise ditch." This prevents oil heated by friction to about 125°F from warming the pipe and melting the permafrost. The elevated portions are cradled by tubular supports placed in the ground and refrigerated to avoid thawing the permafrost. Teflon-coated "shoes" allow the pipe to slide freely within the supports—necessary in case of earthquakes or to accommodate the normal expansion and contraction caused by fluctuating temperatures.

Where the pipeline crosses traditional caribou migration routes, it goes underground. Insulation and a system of small, separate lines carrying refrigerated brine keep the ground around the pipe frozen.

Engineers designed special gate valves at 62 river crossings and other environmentally sensitive areas to automatically shut off the oil flow if the pipe ruptures.

When crossing streams and rivers, the pipeline had to be buried and weighted down with concrete jackets or nine-ton blocks called "saddles." But on 13 crossings, including the 2,300-foot-wide Yukon River, bridges were built to carry the pipe. Since its completion in June 1977, the Trans-Alaska Pipeline has delivered more than nine billion barrels of oil to Valdez.

HEIGHT AND DEPTH

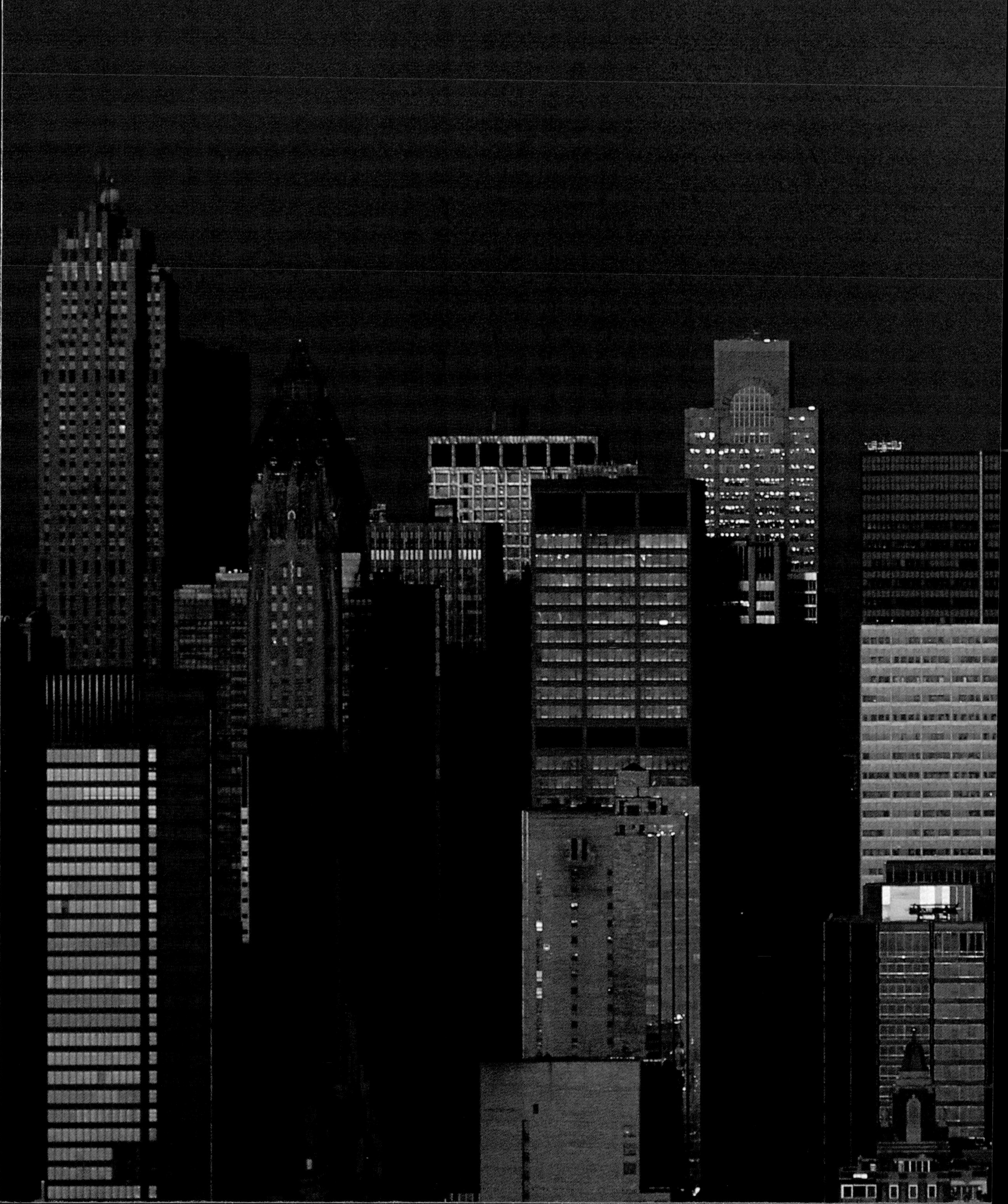

TOWERS
TUNNELS
SKYSCRAPERS

The catalyst for the remarkable feats of construction of the past 200 years was the industrial revolution and its new uses of iron. For the first time, science began to be rigorously applied to the process of construction. As scientific discoveries were made, structural techniques advanced. Bold experiments with iron on long-span bridges in the early 19th century paved the way for Gustave Eiffel's famous 1889 tower, which conclusively demonstrated the capacity of metal to rise higher than pure masonry. Other innovations included Elisha Otis's elevator, techniques of mass production, and new methods of prefabrication, many of which had been put to use in the Crystal Palace of 1851.

Engineers applied these lessons to buildings, relying first on iron and, by the 1880s, on steel to support increasingly taller structures. The first uses of iron-and-steel frames by William Le Baron Jenney and others in Chicago signaled the beginning of skyscrapers. Reinforced concrete, a European invention patented in 1867, also had a profound impact on buildings.

During the 19th century, the job of the engineer grew more specialized, finally emerging as a profession distinct from that of architect. At first, engineers were thought of as suppliers of materials, while architects were credited with the structure's form. But as it became seen as both honest and

JOHN HANCOCK TOWER, BOSTON

THE GATEWAY ARCH, ST. LOUIS

THE CHANNEL TUNNEL, ENGLAND–FRANCE

PRECEDING PAGES: Sun glances off the aluminum cladding of New York's 59-story Citicorp Center in midtown Manhattan. Its distinctive, bladelike roof was originally designed to hold solar panels, but has never been used for that purpose. Completed in 1977, the 919-foot Citicorp Center has a "tuned mass damper," the latest technology to counteract building oscillation caused by wind. In the crown, a 400-ton block of concrete sits on a movable platform connected to the building by large hydraulic arms. Whenever the tower sways more than one foot a second, a computer directs the arms to move the block in the other direction. This action dampens such oscillation by 40 percent, considerably easing the discomfort of the building's occupants during high winds.

THE BANK OF CHINA, HONG KONG

aesthetically desirable to display a building's structural qualities, the engineer's status went up.

After World War II, engineers applied the growing body of knowledge about structural mechanics to develop internal frames and bracing so strong that entire buildings could be covered only in glass. In the 1970s and 1980s, engineers introduced rigid-tube construction, which enabled buildings to rise 100 stories and higher. At the same time, Toronto's CN Tower was built in prestressed concrete and steel, the world's tallest unsupported structure.

Today, computer technology has wrought another revolution. Computers generate theoretical structures for study and selection long before the first hole is dug. Some sophisticated computer programs perform structural analysis, while others calculate the impact of wind shear on scale models in wind tunnels. Still others gauge the impact of earthquake shock. Future technological breakthroughs will include the use of construction robots.

The computer revolution has also had an impact on tunnel building. Engineers working on the Channel Tunnel used a computer and laser system, aided by four Navstar satellites, to continually monitor both ends of the tunnel as they were dug. Whenever the tunnel moved off course even slightly, the computer automatically signaled a course correction to hydraulic jacks, which adjusted the direction of the excavation machinery.

Clues to the future of tall buildings may lie in such structures as the 1,209-foot Bank of China in Hong Kong. It blends steel and concrete in an innovative way, creating a very rigid, light, and economical alternative to traditional designs. Instead of the standard internal steel frame, the builders used a space-frame design of three-dimensional trussing. Perhaps Frank Lloyd Wright's dream of a "mile-high" skyscraper may yet one day be realized.

OFFSHORE OIL RIG, UNITED STATES

Towers

Canted at a dizzying angle, Italy's Leaning Tower of Pisa has listed 17 feet off true vertical during the past 700 years.

F or centuries, towers have exerted a powerful, often spiritual, influence on those who view them. In the Bible, the Tower of Babel symbolized mankind's audacity and presumed self-sufficiency. When it rose too close to heaven, God struck it down.

Builders erected early masonry towers for military or religious purposes; the medieval campaniles, such as that of St. Mark's in Venice, are notable examples. Lighthouses were also early masonry tower forms; the ancient Pharos of Alexandria stood in the Mediterranean until 1326.

Tower building changed dramatically during the industrial revolution. Strong building materials, such as iron, became available, enabling builders to push structures higher into the sky. New concepts in designing foundations and prefabricating sections were among a number of engineering innovations. No longer in service of religion or war, towers became symbolic of a new age of technology and material success. Countries across the world proudly erected towers purely as engineering feats, often as tributes to national heroes.

Since then, newer materials, such as steel and reinforced concrete, have replaced iron. Gasoline-powered cranes have superseded steam-powered ones. Beginning with the designers of the 1889 Eiffel Tower and the previous record holder in height, the 1884 Washington Monument,

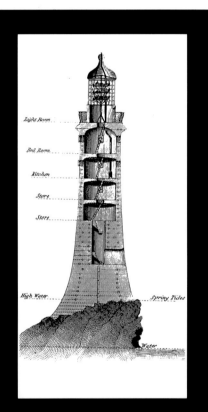

John Smeaton (1724-1792), England's first "civil engineer," built the Eddystone Light (above) on a rocky islet off southwest England. Though not the first lighthouse here, Smeaton's followed a unique design using iron-reinforced granite and waterproof cement. It lasted a record 123 years.

successful builders of tall towers have followed at least two basic engineering principles: the need for a strong foundation and a strategy to resist the destructive force of the wind.

The tilt of one of the world's most famous towers, the Leaning Tower of Pisa, grew out of a fundamental error made by its builders. In constructing the tower's foundation, one not much wider than the diameter of the tower itself, 12th-century Pisan engineers concentrated an excessive amount of weight in a small area of soft soil. Over 700 years, the soil has settled unevenly under the weight of the tower, causing it to lean.

For this reason, tower engineers spend much time and effort in first building a foundation. Because rock surfaces are often not the sites selected for towers, engineers have had to dig to bedrock, sink piles or caissons, or spread the load over a wider area.

Even with a strong foundation, however, a tall tower can topple in a strong wind. Relatively short masonry towers resist the wind through the deadweight of stone. Wind becomes the largest single concern of engineers building newer lighter and taller towers. For his famous tower, Gustave Eiffel designed an open latticework structure with a new form to withstand wind stress efficiently.

The Eiffel Tower

From a distance, the Eiffel Tower seems poised for flight—a sleek rocket ship out of Jules Verne. Up close, the lines turn into a dizzying web of intersecting iron girders.

The simple yet masterful 9,700-ton iron structure rises nearly 1,000 feet into the Parisian sky. Completed in 1889 to commemorate the centennial of the French Revolution, the Eiffel Tower soared nearly twice as high as the world's then tallest man-made structure, the Washington Monument. A testament to French technical proficiency, it remained the world's tallest structure for 40 years.

The genius behind the tower was Gustave Eiffel, a celebrated structural engineer and bridge builder. Iron was the material of the industrial revolution, and Eiffel was a master of the medium. His plan for a tower was selected from 700 other designs.

For the first time in history, abstract engineering forms were the goal of a tower builder. Before then, function was dictated by human concerns: worship or military surveillance. Instead, Eiffel defined function as the maximum load-bearing capacity of the tower. He designed the graceful tapering lines of the piers to resist the force of

Thousands of lights call attention to the elegance and structural simplicity of France's most famous landmark, the Eiffel Tower.

Built on a bank of the Seine, the tower's foundations demanded engineering used in bridge building, one of Gustave Eiffel's specialties. Because water infiltration made surface digging difficult, he sank metallic compressed-air caissons, not unlike diving bells, into the soft soil. Inside them, workers dug out the foundations.

Building the Eiffel Tower

Archival photographs from 1888, when most of the ironwork was erected, reveal the construction process (above and below). Horse-drawn drays brought prefabricated sections to the base of the structure from Eiffel's workshop three miles away. Steam-powered cranes then hoisted the sections into position.

With classic French insouciance, a workman (above, right) paints a column in 1953, balancing nearly 900 feet above the Champ de Mars. Every seven years the tower receives a new coat of muddy brown paint.

4. 27 avril 1888.

3. 7 janvier 1888.

the wind. In his design, Eiffel redefined the relationship between form and function, creating an engineering art form.

Although engineers of the time had extensive experience using iron to build long span bridges, there were few precedents for high structures. But Eiffel had been studying the problem of tower design since the 1860s, when he began to build high railroad viaducts through the windy valleys of the Massif Central.

He tackled the engineering problem with a combination of mathematics, extreme precision, and patience, employing 30 draftsmen for 18 months. Because the edges of the tower were curved and the

trusses and supports were graduated from top to bottom, each piece was designed independently, to reflect variable inclinations and to bear different loads.

Time constraints and the sheer magnitude of the project motivated Eiffel to prefabricate all the sections off-site, a novel development that Eiffel had used for his bridges. Precision played an essential role because the wrought iron pieces had to be riveted together. Rivet holes had to match exactly where placed. All told, some two and a half million rivets lock 12,000 pieces together. Eiffel's care paid off: Even when cranes raised a section into place 164 feet up, the rivet holes matched.

No amount of precision, however, could take into account all the slight variations. To accommodate adjustments, Eiffel devised hydraulic jacks that fit inside each of the four main columns composing the structure's legs. Using the jacks, each capable of lifting 900 tons, Eiffel could minutely adjust the angle of the main columns—a spectacular feat for the era.

Two prefabricated iron sections, pulled up to working level by 50-ton mobile cranes, were then riveted into place by hand. The weight of one crane counterbalanced the other.

Hydraulic jacks placed inside the base of each leg were characteristic of Eiffel's engineering brilliance. By raising or lowering the jacks, the upper elements of the structure could be brought into perfect alignment.

French engineer Gustave Eiffel (1832-1923) worked with iron to create engineering masterpieces the world over. He designed the iron skeleton of the Statue of Liberty, the 1,667-foot-long Garabit Viaduct in central France, and many other structures. His crowning achievement, the Eiffel Tower, was the first truly large-scale industrialized construction project. To it he brought his knowledge of the strength limits of iron, of the effect of wind forces on tall bridges, and of how to set strong foundations in soft soil.

A filigree of steel, the tower's surface is composed of light trusses (left).

FOLLOWING PAGES: Modern double-decked elevators carry tourists at some 2.4 feet per second to a restaurant and an observation deck.

With its simple, elegant lines, the Washington Monument serves as the triumphant centerpiece of the Mall, Washington, D.C.'s grassy promenade. Honoring George Washington, it rises 555 feet—the tallest masonry tower in the world.

Thomas Casey, an Army engineer famous for his Civil War forts, inherited the project in 1876, after work on it had stalled at the 156-foot mark for two decades.

Casey had first to reinforce an inadequate foundation, a challenge made difficult by the fact that about one-quarter of the monument was already standing. Without displacing earth that supported the structure, he carefully removed large amounts of soil from underneath and around the foundation, immediately filling the holes with concrete. His plan utilized a

Insuring against a lightning strike, a worker checks the rods surrounding the monument's solid aluminum tip during the first complete cleaning, in 1935. Planners selected pure aluminum because it is highly conductive and does not tarnish.

White marble quarried in Maryland faces the Washington Monument, the world's tallest structure until the Eiffel Tower was built.

The Washington Monument

For two decades, starting in 1854, the unfinished monument (bottom) stood abandoned, a bleak reminder of Civil War era turmoil.

Reinforcing a weak foundation—one that weighed 32,500 tons—was Thomas Casey's first task in 1876, and it earned him worldwide acclaim. His means, though not new, had never before been tried on so large a scale.

network of iron rail tracks, derricks, concrete mixers, and hoisting equipment. Concrete buttresses added additional support and also served to connect the old and new foundations.

Casey's upper walls are significantly thinner than the earlier, lower ones, reflecting his professional skill. To place them, he devised an iron skeleton for the monument's interior. This acted as a support for a temporary platform elevator to carry stonework. The skeleton also anchored four derricks that swung the masonry into place.

For the 300-ton pyramidal roof, rising the last 55 feet, marble slabs up to 7 inches thick were laid on 12 vertical ribs that converged at the apex. The final piece, the 3,300-pound capstone, was set in 1884, 5 decades after the project had begun.

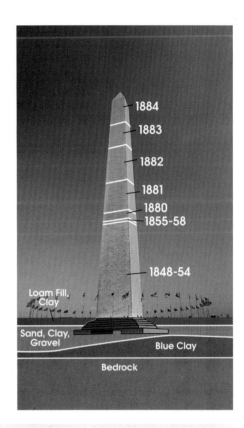

1884
1883
1882
1881
1880
1855-58
1848-54

Loam Fill, Clay
Sand, Clay, Gravel
Blue Clay
Bedrock

The Gateway Arch

Though the arch has been a common architectural feature since ancient times, it reached its zenith in 1965 with the completion of St. Louis's Gateway Arch. The nation's tallest monument, it soars 630 feet into the sky. The marvel is that it stands at all.

In 1948, Finnish-American architect Eero Saarinen won the design competition for the monument, beating 171 other entrants. But his plans for a self-supporting, stainless steel arch the size of a 60-story building were soon shown to be fatally flawed: Two individual legs that curved inward to join at the center would be unstable at such a height. More than a decade later, project engineers led by John Dinkeloo, who had worked with Saarinen, devised a new engineering plan based on a single hollow curving tube; they looked to a special building material and deep foundations to stabilize the form.

Stainless steel was chosen for its tensile strength, non-corrosiveness, and beauty. The Pittsburgh-Des Moines Steel Company fabricated the sections for the arch out of 886 tons of the material—more stainless steel than has been used on any other single construction project. The cost was exceptionally high.

The arch features a strong, stressed-skin design. A cross section of one of the legs reveals an equilateral triangle; each of the triangle's three sides is built from two walls of steel plate. High-strength steel bolts connect the two plates. Concrete, bolstered by steel tensioning rods, fills the space between the plates. By tightening the bolts and rods, the steel skin was stretched tight.

Each leg tapers from 54 feet on a side at ground level to 17 feet at the apex. The resulting structure supports itself.

The self-supporting Gateway Arch (opposite) relies partly on a super-strong steel skin and partly on a solid filling of concrete that encases highly stressed steel rods, carried deep into the arch's foundations.

Looking west, the 630-foot-high arch frames much of downtown St. Louis. Both the structure and the 82-acre Jefferson National Expansion Memorial it straddles pay tribute to the thousands of pioneers who journeyed westward from St. Louis.

The Gateway Arch

The construction team lifted the first six steel sections into place with standard ground-based cranes. After that, two special cranes placed on mobile derricks ("creeper cranes") were used. As work progressed, each crane climbed higher on a track system attached directly to the leg. Once each section of the arch was raised, it was bolted into place.

On October 28, 1965, as jacks pushed the two legs apart and hundreds of people watched, workers dropped the keystone section into place. The creeper cranes then backed down the legs, taking up their tracks as they went.

A fish-eye lens captures a creeper crane as it pulls a steel section skyward for placement 500 feet above the Mississippi riverbank.

An excavation (left), nearly half of it through bedrock, lays the groundwork for the foundation of one of the arch's legs and the underground visitors' center.

Anchored to vertical tracks mounted on each leg, huge creeper cranes (right)—each measuring 43 by 32 feet—act as workstations in the air, hoisting each triangular steel section from the ground to its place on the arch.

A temporary, 58-ton stabilizing strut connects the two legs in preparation for placement of the final sections. Until the legs reached 530 feet, their strength was in equilibrium, and they could free stand. But beyond that, the construction load, including the creeper cranes, made the structure top-heavy. The strut—held by jacks and a steel harness—allowed the weight to be transmitted to the foundations.

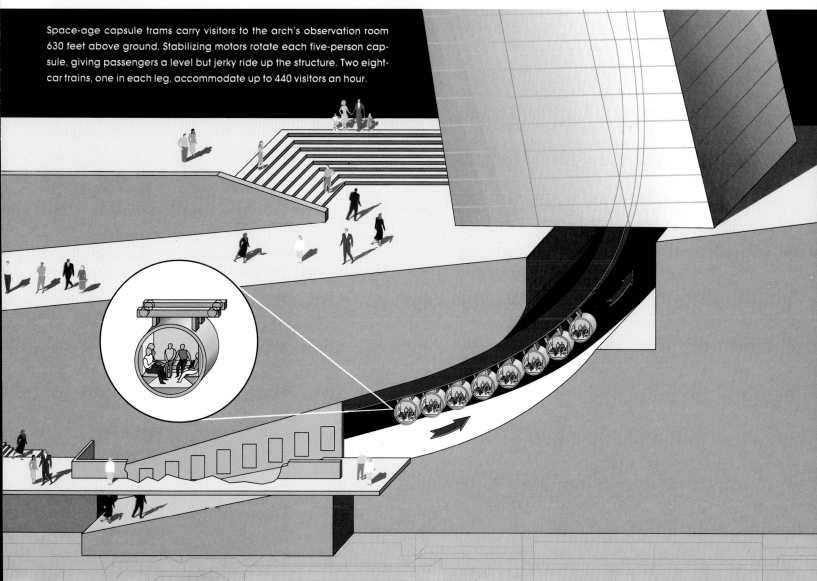

Space-age capsule trams carry visitors to the arch's observation room 630 feet above ground. Stabilizing motors rotate each five-person capsule, giving passengers a level but jerky ride up the structure. Two eight-car trains, one in each leg, accommodate up to 440 visitors an hour.

The CN Tower

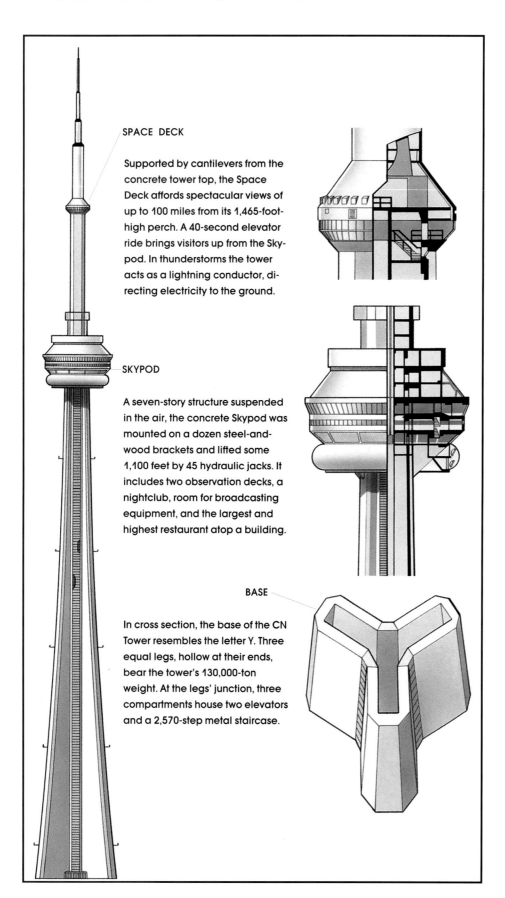

SPACE DECK

Supported by cantilevers from the concrete tower top, the Space Deck affords spectacular views of up to 100 miles from its 1,465-foot-high perch. A 40-second elevator ride brings visitors up from the Skypod. In thunderstorms the tower acts as a lightning conductor, directing electricity to the ground.

SKYPOD

A seven-story structure suspended in the air, the concrete Skypod was mounted on a dozen steel-and-wood brackets and lifted some 1,100 feet by 45 hydraulic jacks. It includes two observation decks, a nightclub, room for broadcasting equipment, and the largest and highest restaurant atop a building.

BASE

In cross section, the base of the CN Tower resembles the letter Y. Three equal legs, hollow at their ends, bear the tower's 130,000-ton weight. At the legs' junction, three compartments house two elevators and a 2,570-step metal staircase.

If the Eiffel Tower looks like a 19th-century rocket ship, then Toronto's CN Tower represents a spaceship of the future. The tallest freestanding structure in the world, the CN Tower rises like a needle 1,815 feet above the city's downtown. Remarkably, it deviates from true vertical by little more than an inch.

Finished in 1975, the CN Tower was built by Canadian National Railways as a means of improving television reception in the area. Gently tapering as it rises, the tower is made of prestressed concrete and reinforced steel. Its thin, narrow profile is deceptive: The tower is so strong that 120-mile-an-hour winds would create a barely discernible wobble at the Skypod level.

Construction took just 40 months, thanks to the ingenious use of a building technique that, though long established, had never before been attempted on such a scale. Mixed on-site, special high-quality concrete was poured into a massive mold, called a slip form, attached to the base. As the concrete hardened, a ring of climbing hydraulic jacks moved the slip form higher. With this technique, the tower rose about 20 feet a day. A helicopter mounted the 335-foot steel transmission mast by airlifting the 39 mast sections into place.

Workers excavated more than 62,000 tons of earth for the structure's massive foundations. They rest on a shale bed 50 feet below ground.

Glass-fronted elevators whisk visitors to the Skypod at a speed of nearly 20 feet per second. Special sensors reduce the speed of the elevators during heavy winds.

Futuristic symbol of the television age, the CN Tower pierces the Toronto sky. Broadcast antennas on the transmission mast are protected from ice buildup by 270 fiberglass panels.

The Statfjord B Platform

Capable of withstanding 100-foot waves and the punishing winds of the North Sea, the 890-foot-high Statfjord B Oil and Gas Platform accommodates 204 people and stores two million barrels of oil, as well as the equipment for recovering up to 250,000 barrels a day.

Statfjord B is the heaviest movable structure ever built, relying on its 890,000-ton weight to anchor it to the seafloor in 489 feet of water. Construction workers completed the massive platform in 1981, when they joined the pre-assembled base and deck in a protected, deep-water Norwegian fjord. Twenty-four reinforced concrete cylinders make up the base; from four of them rise hollow concrete legs (the other 20 store crude oil). On top of the legs sits a steel deck, which holds equipment, a hotel, and a helipad. Joining the 40,000-ton deck and the base was a delicate operation requiring enormous precision.

Tugboats towed the finished platform to the Statfjord field, the world's largest offshore oil and gas field, 114 miles west of Norway. There, engineers filled Statfjord B's 20 concrete base cylinders with seawater, sinking the platform to the seafloor. A steel skirt built around the cylinders penetrated 13 feet into the floor, adding stability.

Drilling machinery operates through two of the four legs. Statfjord B's oil recovery proved such a success that it recouped the $1.8 billion construction costs for its multinational owners before celebrating its first anniversary at sea.

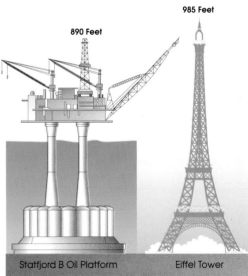

890 Feet

985 Feet

Statfjord B Oil Platform   Eiffel Tower

Its 40,000-ton steel deck and base successfully mated, Statfjord B (opposite) awaits transport from a Norwegian fjord to the North Sea's Statfjord field. The integrated platform's drilling rig, helipad, and living areas are clearly visible.

Engineers used slip-form technology to build Statfjord B's four reinforced concrete legs (left). Four concrete crews had to work simultaneously to maintain the platform's center of gravity.

Tunnels

A work crew files through a tube of the Channel Tunnel, the 31-mile-long passage connecting France and Great Britain.

USING A DIAMOND DRILL TO BORE THE MONT CENIS TUNNEL.

Ever since man learned to enlarge the cave, he has found a number of reasons to tunnel through the earth: for mining and storage, for carrying water, for sewerage and drainage, and for transportation.

Commonly, a tunnel is a horizontal underground passage excavated, or advanced, from the inside and lined to support the surrounding ground. It may bore through low ground or high, passing under city streets, beneath a riverbed or seafloor, or through the side of a mountain. In cross section, its walls can be circular, oval, or a horseshoe, shapes that best withstand earth pressure. Like other large structures, most tunnels rely on foundations to transfer massive loads into the earth below.

Engineers determine a tunnel's form and building method by its function and the type of ground through which it will go, whether hard or soft. Once excavated, soft ground must be reinforced; rock tunnels often also need support.

Today, soft-ground tunnels are normally built with giant tunnel boring machines (TBMs), which excavate, dispose of earth, and add reinforcement. Though rock tunnels have usually been constructed by deep drilling and explosives, new-model TBMs can also bore through some kinds of rock. Shallow tunnels—those under streets—are often excavated from the surface, rather than bored underground.

Mountain Tunnels

Solid rock offers the greatest challenge to tunnel builders and their equipment, whether holing through a mountain or burrowing through a hillside. Piercing hard rock often takes exact geological knowledge and highly specialized technology, both of which have evolved over time.

When test boring along a route isn't possible, as is often the case with mountain tunnels, surveyors must rely on deduction based on surface indications—and a dose of luck. No one can know for sure the nature of the ground inside—or predict when a subterranean pocket of water will burst through a fissure, or when powerful geological forces will damage tunnel supports and explode a rock wall.

The coming of railroads in the 19th century also ushered in the great age of tunnel building. Steep grades prevented trains from traversing high mountains, so engineers had to invent ways to bore through them efficiently and with reasonable safety. In doing so, alpine engineers

A steam locomotive emerges from the extravagantly arched Italian portal of the Mont Cenis, or Frejus, Tunnel (below, left). First tunnel through the Alps, it opened in 1871.

Crew members at Mont Cenis (below) bore a round of holes with air-driven rock drills mounted on a 12-ton carriage. Most of the 80 holes were then loaded with gunpowder and fired in four groups.

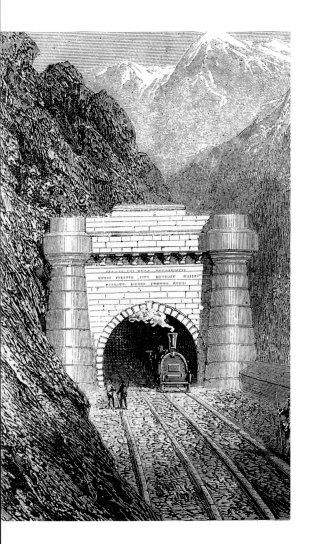

Using carriage-mounted pneumatic drills, workers advance a heading of the Hoosac Tunnel, America's first major rock tunnel, begun in 1851. Behind them, crews enlarge the tunnel with chisels.

had to overcome popular notions such as that high temperatures deep inside a mountain would boil blood. In fact, engineers could not anticipate the real difficulties that lay ahead, including the high costs in money and human life.

For centuries, tunnel builders used hammers and chisels or "fire setting" (the rock face was heated, then doused with water to shatter it) to tunnel through rock.

In the 1600s, gunpowder began to come into use to break rock, and as late as the mid-19th century, blasting gangs still tamped ordinary black gunpowder into holes bored by hand-held steel drills. Typically, the workers drilled blast holes using an ancient technique in which one man hammered a "jumper"—a steel drill with a chisel-shaped bit—while another man held and turned it.

In the early 1850s, work started on an unprecedented five-mile railroad tunnel through Hoosac Mountain in western Massachusetts. At first, the centuries-old methods were used. But by the time the tunnel was completed in 1875, Hoosac engineers were utilizing the type of technology—mechanized drills and reliable explosives—

Mountain Tunnels

that still serves today. Their first attempts to drill included an unwieldy, 70-ton, steam-driven drilling machine. A boiler plant outside the tunnel supplied the steam, which dissipated most of its energy en route to the drills and drenched the workers in hot steam from the drills' exhaust.

In the 1860s, Germaine Sommeiller, the chief engineer of the Mont Cenis, or Frejus, Tunnel—the first such Alpine passage—introduced several major innovations. His "water-spout machine" was an efficient mechanical drill driven by hydraulically compressed air. To maneuver the

machines, he invented a rail-mounted, wheeled carriage that could carry up to nine rock drills and required a crew of 30.

Sommeiller refused to share his drill design with the Hoosac builders, so engineer Charles Burleigh in 1865 devised his own—an even more reliable and powerful pneumatic drill, a hybrid of the steam and compressed-air machines. Burleigh's new drill became the prototype for all piston-type drills that followed.

The year 1866 also saw a breakthrough in explosives with the introduction of nitroglycerine, first employed by the

One of the world's longest highway tunnels pierces Mont Blanc (opposite), western Europe's crown peak. The 31-foot-high tube offers a direct route between Paris and Rome.

Louis Favre, a Genoa builder, won a contract from the Swiss Central Railway Company for a tunnel under St. Gotthard Pass, by guaranteeing to finish it in eight years. If he failed, he would forfeit a huge sum of money. Despite new technology—drilling machines, air compressors, dynamite—the project was doomed from its 1872 start. Favre erred badly in his choice of method, resulting in catastrophic financial losses—and 310 deaths. He died of a heart attack in 1879, three years before completion.

A construction worker (opposite) surveys a stretch of the lined but yet unsurfaced Mont Blanc Tunnel in 1964. Crews gouged enough rock from the mountain to build a sidewalk from Paris to Baghdad.

At the start of their shift, workers push a rail car toward the entrance to Switzerland's St. Gotthard Tunnel in a contemporary engraving. Engineers cut through more than nine miles of solid rock to complete the railroad tunnel in 1882. An auto tunnel opened nearby in 1980.

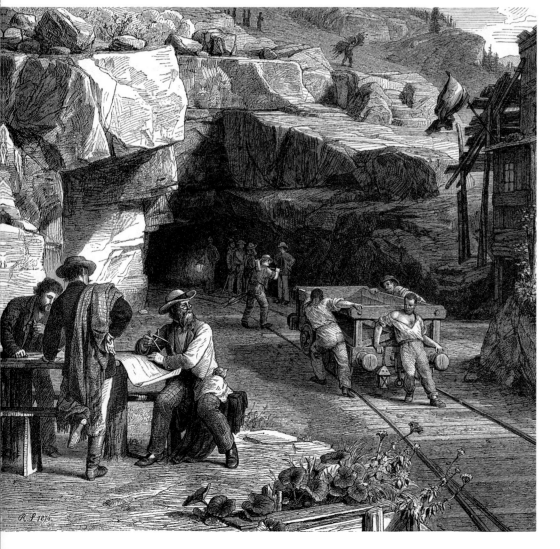

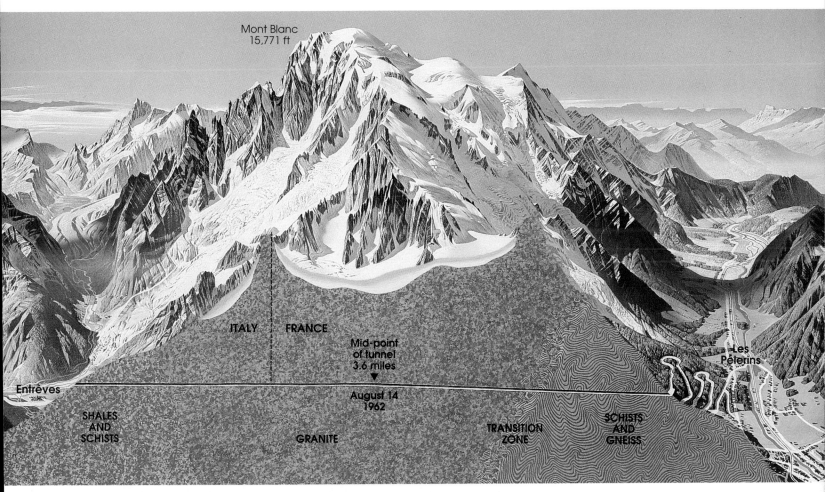

Mont Blanc
15,771 ft

ITALY   FRANCE

Mid-point
of tunnel
3.6 miles
▼

August 14
1962

Entrèves

Les
Péterins

SHALES
AND
SCHISTS

GRANITE

TRANSITION
ZONE

SCHISTS
AND
GNEISS

Hoosac's chief engineer, Thomas Doane. More powerful and less noxious than the traditional blasting powder, though hazardous to handle, the electrically fired explosive could blast far deeper holes and thereby shatter more rock.

Dynamite, the new wonder explosive perfected and patented by Alfred Nobel in 1867, eventually replaced nitroglycerine as the agent of choice. Three years after manufacturing began, it was used to blast a second railroad tunnel more than nine miles through the Alps, at St. Gotthard.

Mountain road tunnels made headlines in 1962 with the breakthrough of the 7.2-mile-long Mont Blanc Tunnel between Italy and France, completed in just three years. Since then, Switzerland has built several even longer auto tunnels.

Underwater Tunnels

ariations in the soft soils of riverbeds and seafloors present special challenges to engineers who build tunnels beneath the water. Whether going through silt, sand, gravel, mud, or clay, such tunnels pose an inherent threat of caving in and flooding.

Tunneling under water was virtually an impossible task before the invention of the protective shield, an ingenious device inspired by the burrowing mole and the teredo, or shipworm—a marine mollusk notorious for boring passageways through submerged timbers.

It was the mechanics of the teredo's strong, protective shell-plates that intrigued British engineer Marc Isambard Brunel. In 1825, he employed a rudimentary tunneling shield modeled after the mollusk's. The shield made possible the construction of

London's Thames Tunnel—the world's first true underwater passage.

Brunel's shield consisted of 12 cast-iron frames, each divided into three vertical floors, or cells. The frames were loosely joined together, forming an 80-ton structure that contained 36 cells measuring about three by six feet each—large enough to hold a worker while he excavated. The shield fit over the tunnel face and temporarily supported it—and the tunnel walls—until the brick lining was in place. The shield was separated into three movable sections advanced by screw jacks.

Over the following decades, the procedure was significantly improved. In 1864, British engineer Peter Barlow patented a one-piece cylindrical shield and a cast-iron lining, built as the shield advanced.

Water breaks through the Thames Tunnel's protective shield on January 12, 1828, drowning six men. Engineer Marc Brunel's son narrowly escaped their fate. Floods occurred five times during the 18-year project.

Planned as a relief for London's traffic, the Thames Tunnel (right) was, in fact, a flop. It remained a curiosity from 1843 until eventually becoming part of the city subway system.

Underwater Tunnels

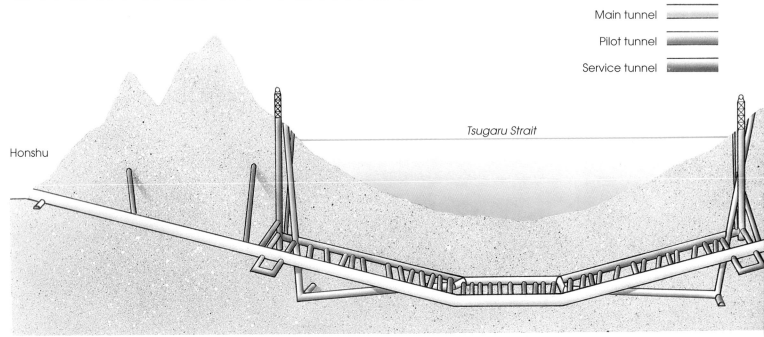

Main tunnel

Pilot tunnel

Service tunnel

Tsugaru Strait

Honshu

The shield allowed tunneling through heavy silt, clay, and mud.

A decade later, Barlow's protégé, James Henry Greathead, devised a cylindrical shield that used compressed air to hold back the water. Greathead's system equalized the air pressure inside the tunnel with the water pressure outside—similar to the pneumatic caisson method used in building underwater bridge foundations. The system was introduced during construction of the City & South London Railway Tunnel in 1886. The twin circular bores measured slightly more than ten feet in diameter and marked the start of modern underwater tunneling.

Since the 1950s, sunken tubes have replaced shield-and-compressed air wherever possible. Initiated as an experiment on the Detroit River Railroad Tunnel in 1906, the procedure calls for sinking sealed, prefabricated sections of tubing into a dredged trench and joining them together, then covering them with backfill. A less expensive system, it also is less risky because air pressure inside the tube

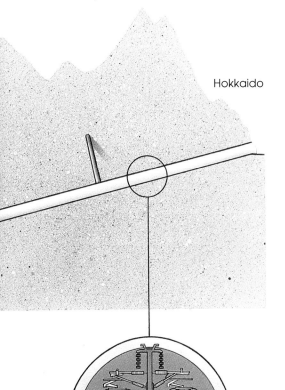

Hokkaido

Narrow   Standard
gauge   gauge

A massive complex of three interconnected tubes forms Japan's 34-mile Seikan Tunnel (above), built between 1971 and 1988. Dug through exceptionally difficult rock, the project took twice as long and cost ten times more than originally expected. Plans for a dual gauge system (inset), with standard-gauge tracks for bullet trains, have been postponed.

An emergency station (opposite) gives direct access to escape tubes through which travelers can be evacuated in case of fire. At either end of the Seikan's submerged section, such stations are essential in so long a tunnel.

is the same as it is at sea level.

In the 1920s, Clifford Holland solved the problem of ventilating automobile tunnels. To remove noxious exhaust gases in the Holland Tunnel, he and his staff devised a system based on powerful electric fans set in ventilation towers on each shore. The fans forced air under the roadway and up into the tunnel through curbside vents. A separate set of exhaust fans sucked the foul air out through ceiling vents. This system changed the air every 90 seconds.

High technology, such as laser-optical surveying and guiding techniques and giant full-area tunnel boring machines—and the use of shotcrete, or sprayed concrete, to stabilize headings and stave off flooding—has revolutionized underwater tunneling.

Ironically, the space age has ushered in a new era of railway tunnels. Notable are Japan's 34-mile-long Seikan Tunnel, the world's longest, completed in 1988, and the Channel Tunnel between Britain and France. Too long to ventilate adequately for motor vehicles, these tunnels are designed for high-speed, nonpolluting electric trains.

"I am going into tunnel work," Clifford M. Holland vowed to a Harvard classmate in 1906, "and I am going to put a lot more into it than I'll ever be paid for." With a proven record on East River subway tunnels, the young Holland in 1919 was appointed chief engineer of an automobile tunnel beneath the Hudson River that would link Manhattan and New Jersey. Holland's obsession with the project ruined his health. He died at age 41 in 1924, never to see his tunnel completed. It opened on November 12, 1927, slightly more than seven years after excavation began.

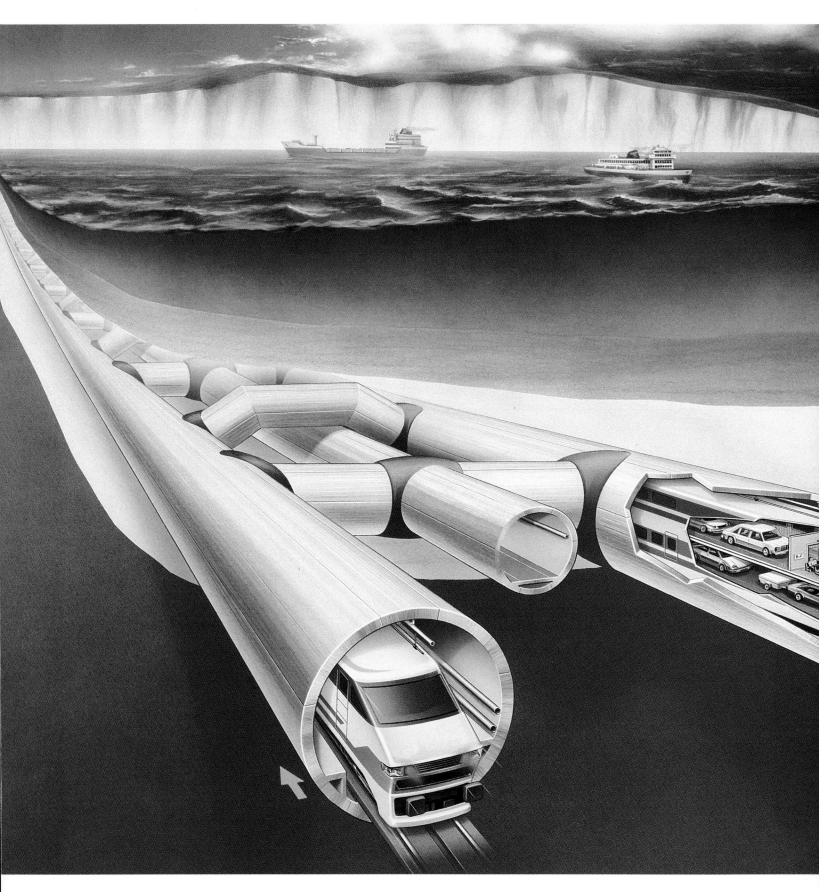

The Channel Tunnel

here are few projects against which there exists a deeper and more enduring prejudice than the construction of a railway tunnel between Dover and Calais," said Winston Churchill in 1936.

Indeed, it has taken two and a half centuries to realize the dream of a trans-Channel link between Great Britain and France. Contractors even started to dig twice, in the early 1880s and again in 1974.

Finally, in 1987, excavation began on the Channel Tunnel, a project to be financed, built, and operated by Eurotunnel, an Anglo-French company. Transmanche Link, an Anglo-French joint venture, contracted for the construction work. On December 1, 1990, crews of both countries shook hands some 200 feet below the seabed in the tunnel breakthrough, uniting the two shores in a fixed link.

The 31-mile-long, three-tunnel "Chunnel" boasts the world's longest underwater section at 24 miles. With building costs likely to total nearly 18 billion dollars, the tunnel is also history's most expensive privately financed engineering project.

Construction involves both ancient and state-of-the-art engineering. Surveyors aligned facing tunnel sections using navigation satellites, triangulation, and plumb

Burrowed into a chalk layer well below the English Channel, the "Chunnel" consists of three parallel tunnels: two one-way rail tunnels for shuttles, freight trains, and high-speed passenger trains, and a central service tunnel, which connects to the main tunnel every 410 yards. The service tunnel gives access for evacuation, maintenance, and fresh air.

Among the fantastic early schemes for a trans-Channel link, Hector Horeau in 1851 proposed a sunken tube straddled by pavilionlike ventilating stations (inset).

Building the Channel Tunnel

lines. Guided by both computer-linked lasers and dead reckoning, technicians some 200 feet below the seabed advanced massive tunnel boring machines, or TBMs—the cutting edge of high-tech tunneling. Traveling from both directions, the TBMs simultaneously lined the tunnels with concrete or cast-iron segments. Ejected chalk marl was transported to the surface for disposal. Near midpoint, the British TBMs nosedived into the Channel floor for permanent burial, enabling the French to complete the breakthrough.

The project posed huge logistical

Workers load concrete tunnel-lining segments, stockpiled like giant ribs, onto a transporter at the Calais precast factory.

Inside the Chunnel, a tunnel boring machine (TBM) dwarfs a construction worker standing before its rotary drilling head. Chalk marl from the excavation cakes the leviathan's tungsten cutting head.

and technical challenges, such as how to transport workers, supplies, and equipment to the headings, and how to remove millions of tons of spoil. The builders had to maneuver huge machines through wet and muck, debug new technologies, and fix jammed conveyors. Despite setbacks, delays, and soaring costs, the Channel Tunnel has earned its place as a landmark of 20th-century civil engineering.

Tracks, used for small diesel engines that shuttle workers and supplies, disappear into one of the two main rail tunnel entrances on the French side (above).

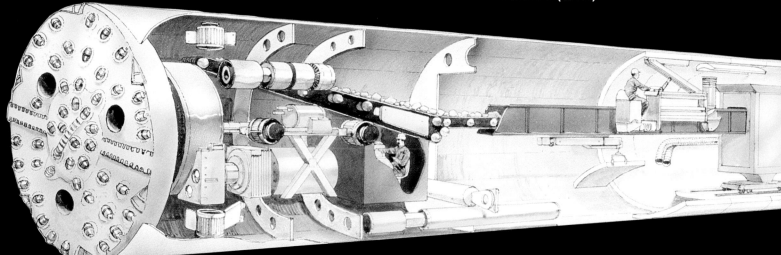

Workers (top) prepare to line the freshly dug tunnel bore. A construction train transports the tunnel-lining segments from access shafts to TBM conveyor belts. Overhead arch the giant hydraulic rams that push the TBM forward. A diagram (above) reveals the anatomy of a TBM, a high-tech assembly line that can simultaneously remove spoil and install tunnel linings.

Skyscrapers

Awash in sunlight, the twin towers of New York's World Trade Center rise 100 stories higher than the earliest skyscrapers.

N o other totems to the modern age are more compelling than the gravity-defying, vertical shafts of glass, steel, and concrete known as skyscrapers. In little more than a century, skyscrapers have proliferated throughout the world, now dominating the skyline of nearly every major city.

While ubiquitous today, the skyscraper is an indigenous American architectural form, which, like most complex innovations, arose from earlier developments. That it emerged in Chicago in the late 19th century is due to several factors, including a booming economy, the rise in value of urban space, and the work of a few brilliant engineers. The devastating fire of 1871 played a large role, giving businessmen the chance to rebuild much of downtown.

Architect William Le Baron Jenney's first Leiter Building, with its open facade of structure and glass, began the evolution in 1879. His Home Insurance Building followed. Though it broke no height records, the ten-story building contained some features that characterized later skyscrapers. Its upper stories were built with Bessemer steel instead of cast iron, and it relied on a steel-and-iron frame, not a masonry wall, to support much of its weight.

Most of the technology necessary for building skyscrapers originated in the early to mid-19th century, a time of large-scale iron projects and the appearance of

Father of the skyscraper, William Le Baron Jenney (1832-1907) studied engineering in Paris and worked as a civil engineer in the Union Army. Using that knowledge as a Chicago architect, he designed the landmark iron-and-steel-framed Home Insurance Building of 1885.

HOME INSURANCE BUILDING, CHICAGO

mass-production techniques. With these developments emerged the profession of structural engineering, stimulated in part by the introduction of iron into buildings and bridges in the late 1700s. As the projects grew larger in the early 19th century, engineers devoted themselves increasingly to the mechanics of the new building materials and to structural analysis. Their experience would result in the development of the steel I-beam, the use of reinforced concrete, and other innovations critical to tall buildings, in which structural integrity is of paramount importance.

The appearance of the safety lift, soon known as the elevator, was another crucial development. Until Elisha Otis's invention in the 1850s, buildings were limited to heights people could walk up comfortably, usually no more than a few stories.

That a strong, internal steel skeleton could now support great weight while remaining light—and that it had reliable tensile strength and a much higher compression strength—encouraged engineers to propose buildings that could soar beyond the limits of traditional masonry structures. Without the need for heavy exterior walls, engineers could now drape non-load-bearing materials onto the frame. Not only did buildings look different—with glass-and-steel shells, marble facades, and fireproof materials—but their occupants

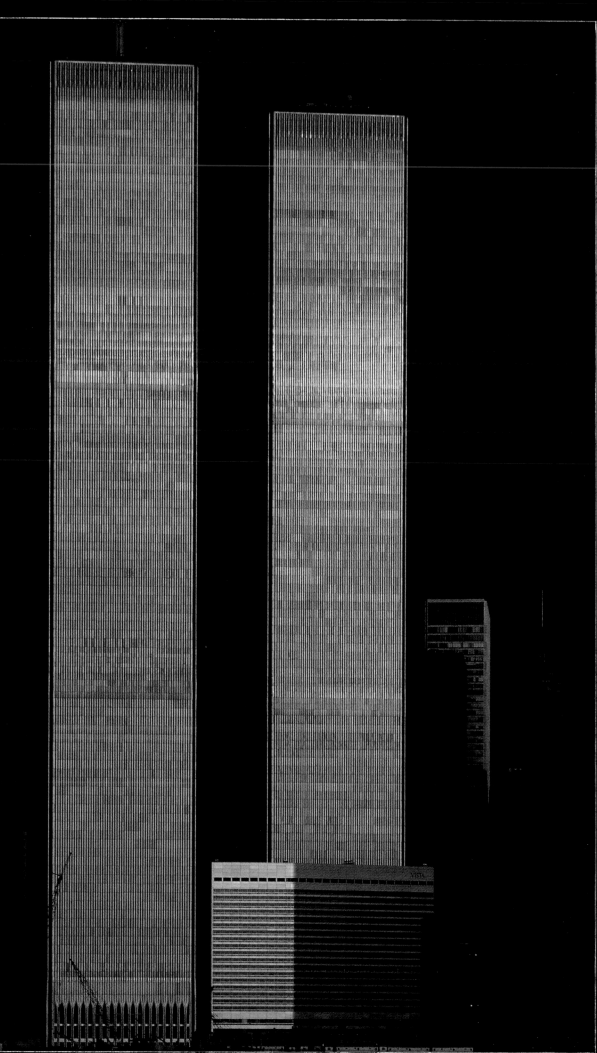

Skyscrapers

benefited from added light, air, and safety.

The steel frame and the diagonal bracing necessary to resist wind pressure were refined during the 1920s and 1930s, preparing the way for the modern period after World War II. In the 1950s and 1960s, many buildings were entirely sheathed in non-load-bearing materials like glass, a technique known as curtain-wall construction. Inexpensive energy for air conditioning was an important prerequisite for these "glass boxes."

Beginning in the early 1960s, many of the most exciting skyscrapers featured tubular design, where walls once again bore loads but now could be light *and* strong.

This design allowed more internal room because it removed, among other things, the need for the diagonal wind braces.

With much of the gravity load and nearly all of the wind load carried by the exterior walls, engineers and architects now began to explore new forms to express structure in skyscrapers. In concrete, this led to buildings that display the smooth transfer of forces from closely spaced wall columns above to large, widely spaced columns at street level, as, for example, in Houston's Two Shell Plaza. In steel, it led to an exterior X-braced tower for the John Hancock Center and to the bundled tube design of the Sears Tower, both in Chicago.

First-generation skyscraper: Chicago's Reliance Building of 1895.

Washington Monument
Washington, D.C., 1885
555 feet
Tallest load-bearing masonry structure in the world.

Montauk Block
Chicago, 1882
130 feet
First ten-story building in Chicago.

Tacoma Building
Chicago, 1889
165 feet
Early use of steel construction with curtain-wall facade.

Auditorium Building
Chicago, 1889
270 feet
Early use of forced-air ventilation.

Rand McNally Building
Chicago, 1890
125 feet
First use of an all-steel frame.

Monadnock Block
Chicago, north half 1891
south half 1893
215 feet
First large building in the U.S. with portal framing.

Reliance Building
Chicago, 1895
200 feet
Riveted steel skeleton with lightweight cladding made it a model for later glass curtain-wall buildings.

Flatiron Building
New York, 1903
286 feet

Ingalls Building
Cincinnati, 1903
210 feet
First reinforced concrete skyscraper.

Woolworth Building
New York, 1913
792 feet
Record height required sinking concrete caissons 100 feet below ground.

Empire State Building
New York, 1931
1,250 feet
World's third tallest building.

RCA Building
New York, 1933
850 feet
First of first large-scale skyscraper complex.

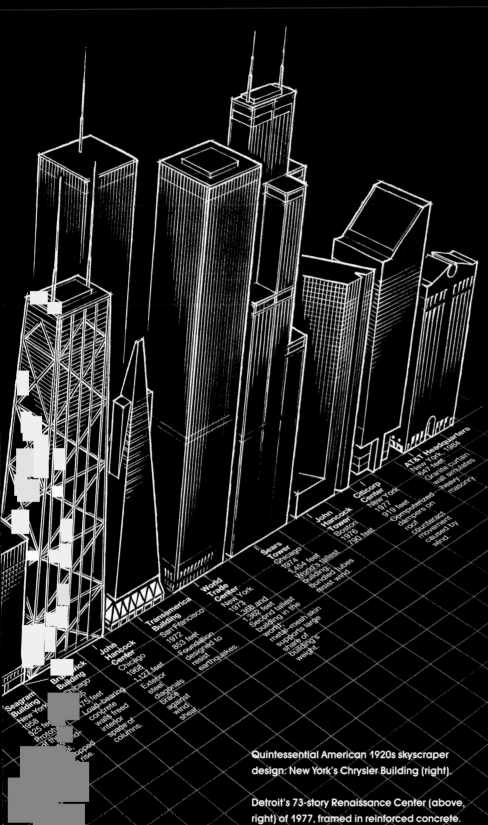

Seagram Building
New York
1958
525 feet
Prototype of the glass and steel–capped rise.

Brunswick Building
Chicago
1965
475 feet
Load-bearing concrete walls freed interior space of columns.

John Hancock Center
Chicago
1968
1,127 feet
Exterior steel diagonals brace against wind sheat.

Transamerica Building
San Francisco
1972
853 feet
Foundation designed to resist earthquakes.

World Trade Center
New York
1973
1,368 and 1,362 feet
Second tallest building in the world; metal-mesh skin supports large share of building's weight.

Sears Tower
Chicago
1974
1,454 feet
World's tallest building; bundled tubes resist wind.

John Hancock Tower
Boston
1976
790 feet

Citicorp Center
New York
1977
919 feet
Computerized dampers on roof counteract movement caused by wind.

AT&T Headquarters
New York, 1984
647 feet
Granite curtain wall simulates heavy masonry.

Quintessential American 1920s skyscraper
design: New York's Chrysler Building (right).

Detroit's 73-story Renaissance Center (above,
right) of 1977, framed in reinforced concrete.

129

Skyscrapers

Downdrafts from New York's 1903 Flatiron Building raised the skirts of female strollers along 23rd Street. To disperse the gawkers, police reputedly shouted "23 Skiddoo!"

The Flatiron Building (opposite), supported entirely by a steel frame, typified the first skyscrapers. Its stone facade is simply decorative.

Though the first skyscrapers built during the 1880s and 1890s reached only modest heights—at 10 to 20 stories, lower than the Washington Monument and the Eiffel Tower—they represented a profound change in building construction. Amid innovations in materials, foundation work, fireproofing, and load-bearing technology, the skyscraper was born.

Beginning with William Le Baron Jenney's Home Insurance Building in 1885, a building's design incorporated an internal steel frame to carry its weight, thus dispensing with the need for large amounts of masonry. Earlier, cast-iron and wrought-iron frames had been used successfully with masonry walls. But it was the emergence of the less expensive steel frame that made tall buildings practical.

Architect John Root, who trained as a civil engineer, freed large buildings from reliance on masonry supports altogether. His 1890 Rand-McNally Building was the first to have an all-steel frame. With partner Daniel Burnham, Root designed nearly 30 major buildings in central Chicago—including the landmark Monadnock Building of 1891, with its unadorned masonry facade that expressed its structure.

Engineer Charles Louis Strobel's innovations in standardizing steel beams led to beams that were both large and long enough to support great weight. His Z-bar column, which looked like an H in cross section, was the precursor of the I-beam used today. The connections between girders, beams, and columns play a

Legions of skeptics warned that Cincinnati's Ingalls Building (left), the world's first reinforced concrete skyscraper, would collapse soon after completion in 1903. It still stands today.

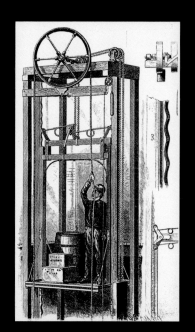

Of all the technical innovations that made skyscrapers possible, perhaps the most important was the safe elevator, invented by Elisha Otis in the early 1850s. His first model, powered by steam, featured an automatic safety device that prevented the car from falling if the hoisting rope broke. Otis installed a passenger elevator in New York in 1857. Hydraulic elevators, which ran on water pressure, were first used commercially in 1878 and are still in use today in low buildings. Not until the advent of electric elevators in the late 1880s, however, did builders deem elevators reliable for tall buildings—or for the Eiffel Tower. Today, elevators in Chicago's John Hancock Center travel its 100 stories at 1,800 feet a minute.

Skyscrapers

crucial role in the overall strength of the steel frame. The elements must be carefully fixed to one another so that a load, such as wind, can be transmitted to the ground by the bending of the beams and columns, rather than by diagonal bracing.

Because the stiff steel frames bore all the vertical and horizontal loads, the walls could now be made of any number of relatively weak materials, such as glass, terracotta, or thin sheets of steel and marble. This resulted in so-called curtain walls: The wall cladding was tacked onto the steel beams. The Tacoma Building of 1889, designed by Holabird and Roche, was one of the first skyscrapers in Chicago to use curtain walls. Their advantages included great

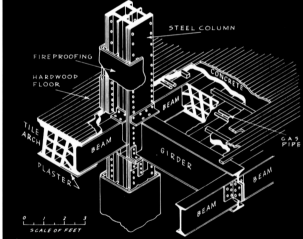

A drawing from William Le Baron Jenney's Chicago Fair Store of 1892 reveals the placement of fireproofing and the riveting that connects beams, girders, and columns. The tile-arch floor would later be replaced by a structural, reinforced concrete slab.

Remarkably modern for 1895, Chicago's 14-story Reliance Building (left), with its open glass walls and horizontal terra-cotta bands, was a direct precursor of the glass-curtain-wall International Style skyscrapers that would predominate in the 1950s and 1960s.

light and air penetration. Their disadvantage was the need for extra wind bracing.

Jenney tackled wind problems by strengthening the steel skeleton with such devices as knee braces—rods placed diagonally across girder and column. These bulky additions, however, interfered with floor plans. A more satisfactory arrangement came with engineer Corydon Purdy's development of portal framing, which thickens the meeting points between girder and column, but leaves the interior space free. Edward Shanklin used portal framing in the 1895 Reliance Building, as did John Root in the Monadnock Building.

Chicago's soft soil and high water table proved a problem for traditional foundations. Root devised a grillage foundation—a raft of rails that spread the weight of each building column over a wide area of ground. The first steel-beam grillage foundation was used in the Tacoma Building. It remains standard today for moderately tall buildings.

The use in buildings of concrete reinforced with metal began in the 1870s. Concrete strongly resists compression and bonds well with metals, protecting them from corrosion and fire. Most important to reinforced concrete's development was French engineer François Hennebique, who propagated his ideas with the slogan *plus d'incendie*—no more fires. By the turn of the century, concrete reinforced with steel was in wide use in buildings, notably the 16-story Ingalls Building in Cincinnati.

To counteract unstable soil conditions, engineers working on New York's 60-story Woolworth Building of 1913 sank caissons into the mud, forced water and dirt out with pneumatic pressure, then filled the caissons with concrete.

Building the Empire State

Its mast originally designed for anchoring zeppelins, the 1,250-foot Empire State Building is now the world's third tallest building.

The building, shown here near its halfway point, consumed enough steel beams to construct a railroad 178 miles long—the distance between New York and Baltimore.

For most people, the history of skyscrapers begins in earnest with the Empire State Building, whose 1,250-foot height and graduated shape help define New York's skyline. Completed well ahead of schedule in 1931, it remained the world's tallest building for four decades, eclipsed only by the 1973 World Trade Center and then by Chicago's 1974 Sears Tower. The Empire State Building rose amazingly fast: Its 85 stories were finished within 18 months.

The gradually recessed walls, or setbacks, were required by New York City's cautious building codes. In 1945, dramatic proof of the skyscraper's stability came when a U.S. Air Force B-25 crashed into the 72nd and 73rd floors, causing damage but no real danger to the overall structure.

The stiff internal skeleton contains 57,000 tons of steel beams, connected by rivets and bolts, and strengthened with portal bracing. Beams were moved to the site by a small-gauge railway and transported skyward by derricks and electric hoists. As many as 38 five-man riveting teams performed a difficult and precarious job of sinking red-hot rivets into pre-drilled holes in the steel. As the rivets cooled, they shrank, solidly fixing the connections.

Daily timetables and progress reports tracked, numbered, and designated steel beams, bricks, window frames, and other building materials. Construction ran with such precision that many steel beams were riveted into place only three days after leaving the Pittsburgh factory.

Legendary photographer Lewis W. Hine captured the precariousness of the work, which killed 14 men in falls and other accidents. To get his shots, Hine would sometimes swing out from the frame in a basket. At top, a workman pounds red-hot rivets into pre-drilled holes to connect steel beam to girder.

A blue-collar acrobat secures a cable one-quarter of a mile above the streets of New York (above, left).

With the Chrysler Building in the background, a bolter finishes part of the Empire State's 57,000-ton steel frame (above).

At a lower level, members of the hoisting gang (left) attach cables to bundles of girders, then give the high sign, sending the steel skyward.

Skyscrapers

Fully exposed steel columns and a separate service tower are among the innovations engineered in Chicago's Inland Steel Building (above), completed in 1958.

Symbol of a new post-war era, the 505-foot United Nations Secretariat (opposite) in 1952 inaugurated New York City's glass-curtain-wall construction boom.

The vertigo-prone need not apply for skyscraper construction jobs: A worker (above, right) installs an antenna cable high atop Chicago's 1,127-foot John Hancock Center.

The modern period of skyscrapers began after World War II, when economics and taste stressed the use of unadorned exteriors, simple lines, and new materials.

Glass sheathed many skyscrapers built during the 1950s. These "glass boxes" were inspired by the work of architect Ludwig Mies van der Rohe, whose International Style in skyscraper building set the tone for the next two decades.

The new modernism refined curtain-wall construction to an art form. Light, pre-assembled, and easy to erect from inside the building, these curtain walls were literally hung on a steel frame. Inexpensive to manufacture in large, even sheets, glass proved a good cladding material. New York's United Nations Secretariat and the more influential Lever House, both completed in 1952, were the city's first glass-curtain-wall buildings. Mies's elegantly styled, 38-story Seagram Building, finished in 1958, also spawned a host of imitators.

While heating and cooling proved more difficult with glass, which is highly conductive, the development of air conditioning and the low cost of energy in the 1950s relieved the problem.

Perhaps the most significant skyscraper of the 1950s, from an engineering standpoint, was Chicago's Inland Steel Building, which featured a stainless-steel curtain wall with steel columns set out from it. A precursor to many modern buildings, Inland Steel had "clear-span" construction: Its 14 exterior columns anchored long girders that supported each floor. Inland Steel was built not on concrete caissons but on less expensive steel pilings.

Though curtain-wall construction proved popular, certain drawbacks caused engineers to look for new designs. Without the additional rigidity of an exterior material such as masonry, curtain-wall buildings needed extra-rigid and more expensive frames to carry gravity and wind loads. Such internal bracing ate up costly floor space.

In the early 1960s, engineers, led by Fazlur Khan and Leslie Robertson, made a breakthrough. If, they reasoned, the exterior walls of a building were treated as a rigid tube, then much internal bracing could be removed. Khan, with architect Bruce Graham, first implemented this idea in Chicago's 1965 DeWitt-Chestnut Building, in

Skyscrapers

Engineer Fazlur Khan's "bundled tube" construction in Chicago's 1,454-foot Sears Tower, the world's tallest building, required much less steel than did traditional designs, while maintaining stiffness against wind loads. Tubes terminate at the 50th, 66th, and 90th floors to allow for floors of different sizes and to give the building its graceful shape.

which the entire concrete exterior forms a single tube that takes gravity loads and all the wind load. The same designers modified the idea for Chicago's 37-story Brunswick Building. Completed in 1965, it features an exterior structure made of reinforced concrete that bears the building's weight and is connected to an interior tube by concrete floors. The tube, housing elevators and other service-related equipment, resists the horizontal wind load. In the early 1960s, Robertson designed the World Trade Center towers as tubes, although they were not completed until 1973.

In the mid-1960s, Fazlur Khan, in-

In the 1960s and '70s, Bangladesh-born engineer Fazlur Khan (1929-1982), who became a partner at Skidmore, Owings and Merrill, collaborated on some of the world's tallest and most innovative skyscrapers. His work defined the Second Chicago School, helping to reshape standard construction approaches. As a master engineer—and a structural artist—Khan said that "a building's natural strength should be expressed." His designs gracefully blend form and function.

spired by architect Myron Goldsmith's idea for a diagonally braced skyscraper, designed the John Hancock Center in Chicago with Bruce Graham. With diagonal steel members each crossing 18 stories, the exterior structure easily carries much of the gravity loads and all the wind load with relatively little steel. The 100-story skyscraper opened in 1970.

The world's tallest building, Chicago's 1974 Sears Tower, relies on another innovative tubular system, known as the "bundled tube" approach. To the height of 50 stories, nine steel-framed tubes, each 75 feet square, are interlocked. Each serves to reinforce the others. At regular intervals, the tubes stop, leaving only two that rise to the 110th floor.

Direct descendant of the Eiffel Tower, Chicago's John Hancock Center (above, right) boldly expresses structure. The diagonals lace the vertical columns together and thus distribute the gravity loads equally among them.

In the aftermath of the devastating 1906 earthquake, San Francisco's downtown was littered with rubble from collapsed buildings. Most wood-framed and steel-framed buildings, however, survived with little structural damage.

Buildings with such frames, engineers realized, were far superior to unreinforced masonry buildings in resisting the strong lateral forces of an earthquake. While clearly all structures must support vertical loads because of their own weight, designers had given little thought to lateral forces, except with regard to wind shear. The San Francisco quake, followed by one in Tokyo in 1923, forced engineers to confront the issue.

An earthquake produces a series of waves that moves across the ground, caus-

ing a building's foundation and superstructure to sway from side to side. Engineers learned that while a rigid building—one of unreinforced masonry—could resist some violent shaking, the best defense against these lateral forces was a building that could sway with the earthquake. By bending with the wave, a structure can absorb much of the wave's destructive energy. A building designed with rubber or metal springs, for instance, would deflect the most violent earthquake. Such structures, however, would not carry the vertical load for even the smallest building.

Steel and reinforced concrete (which was not yet in wide use in San Francisco at the time of the quake) are the best answers to balancing flexibility with

Building to Withstand Earthquakes

The massive earthquake of 1906 leveled many masonry buildings in San Francisco and such neighboring towns as Santa Rosa (opposite).

The pyramid shape of San Francisco's Trans-america Building (opposite, lower), and a heavy mat foundation, help it resist earthquakes.

A building's configuration, design, and location are uppermost in determining whether and how a building will be damaged by an earthquake (below). A structure whose long axis is built parallel to the ground motion (1) will sway far less than one built perpendicular to it. If dissimilar structures are joined (2), they may rupture at the connection. Similarly, the connection between a building's tower and lower portion (3) may crack.

A building with a "soft" first story, composed only of tall columns (4B), is vulnerable because it can't distribute the force of the earthquake throughout its entire form as another building can (4A). The force of an earthquake on a building that sways at the same frequency as the earthquake (5) will be intense and probably destructive. Buildings near one another (6) may collide during an earthquake.

FOLLOWING PAGES: In the 1980s, Hong Kong added high technology to its crowded, cement-and-glass-box skyline with such buildings as the Hongkong Bank and the Bank of China, symbols of its aspirations for the future.

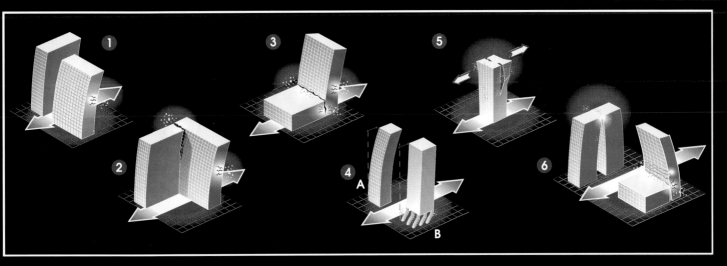

load-bearing capacity. In addition, so-called "passive solutions," incorporated into building design, help absorb the impact of earthquakes. In the base isolation technique, alternating layers of rubber and steel plates are placed between the base of the building and its foundation. These dampen shock waves and thus reduce the effects of a quake. Other passive methods include the installation of sliding bearings under a building's load-bearing columns to reduce the earthquake forces and dissipate the energy through friction.

In addition, engineers have focused on the problem of soil liquefaction, the sudden breaking up during an earthquake of loose and wet soil beneath a building. One solution is to place steel-reinforced piles below the building, which transmit the building loads through weak soil to firmer soil at greater depth.

Recently, American and Japanese engineers have launched a revolution in so-called "active solutions," in which sensors detect and predict earthquake waves, then signal computer-controlled devices to counteract the effects of the earthquake. In an experimental building in Tokyo, large weights in its upper levels swing in the opposite direction of the shock wave, damping the motion. (New York's Citicorp Center uses a similar system to counteract wind shear.) Such active systems require emergency back-up power to keep them operating in the event of a power loss during an earthquake.

A shaking table at UC Berkeley's Earthquake Engineering Research Center tests the damping effect of rubber and steel plates.

The Hong Kong Banks

F ew cities have the concentration of sky-scrapers—or banks—that Hong Kong has. Two rival banks completed in the late 1980s, and located downtown two blocks apart, have drawn particular attention: one, the Hongkong and Shanghai Bank Headquarters, for its dramatic, high-tech appearance; the other, the Bank of China, for its highly innovative structure. The architects and engineers of both banks faced many challenges, including Hong Kong's crowded cityscape and its fierce winds.

The premium on office space prompted the design of the 43-story Hongkong Bank. Its engineering and mechanical features are completely visible on the north and south facades. Eight masts, lined up along the perimeter of the building, carry its weight, resulting in an open floor plan. Each mast weighs more than 1,000 tons and contains four steel tubes

connected by crossbeams. At five points, large suspension trusses form two-story triangles, which carry the loads of the floors to the masts.

Much more economical was the 70-story Bank of China, the tallest skyscraper outside the United States—and to many people, the finest since Mies van der Rohe's Seagram Building of 1958. Architect I.M. Pei and engineer Leslie Robertson relied on a "megastructure"—a triangular trussed grid of glass and metal—instead of the usual internal steel frame. No internal diagonal wind bracing was required, and the design called for 40 percent less steel

Built by architect Norman Foster and the engineering firm of Ove Arup & Partners, the Hongkong Bank's floors hang from five massive, aluminum-clad suspension trusses.

Computerized, mirrorlike "sunscoops" fill the Hongkong Bank's atrium with light (opposite).

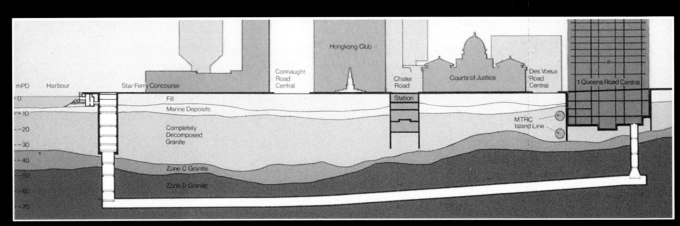

The waterlogged earth beneath the site of the Hongkong Bank presented a problem. Draining it, studies showed, might destabilize nearby buildings. Instead, laborers dug trenches and lifted out rubble until reaching bedrock; then they poured concrete through pipes, enclosing the area in a huge wall. Crews then built a basement from the top down, and from it excavated a tunnel some 250 feet below the street—to avoid the city's utilities and underground railroad. Through the tunnel, intake, outtake, and reserve pipes (left) carry water 1,150 feet from the harbor for air conditioning and flushing .

than is usual in buildings of such height.

The Bank of China starts as a 165-foot-square shaft; two diagonals divide it into four triangular quadrants. The first, second, and third quadrants disappear at various stages as the building rises, leaving one thin tower at the top. Five main columns hold up the building; at its four corners, all steel members are joined in reinforced concrete blocks that add to the structure's stability and wind resistance.

If the building has a liability, it is that triangles augur danger, according to *feng shui*—the Chinese art of building in harmony with nature—causing anxiety in the area.

Showpiece of the Beijing government in Hong Kong, the 1,209-foot, glass-and-aluminum-clad Bank of China dominates the cityscape.

SPORTS ARENAS
EXPOSITION HALLS

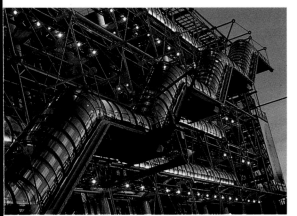

POMPIDOU CENTER, PARIS, FRANCE

GUM DEPARTMENT STORE, MOSCOW, RUSSIA

Over the ages, great ingenuity has been devoted to creating functional buildings for large-scale public entertainments. As many as 55,000 spectators in ancient Rome's huge Colosseum cheered gladiators from beneath a retractable canvas awning whose supports, anchored in the exterior walls, allowed people an unimpeded view. Today, 150,000 soccer fans applaud their teams from well-planned seating in Rio de Janeiro's Maracanã Stadium, with its roof of reinforced concrete.

The development of coverings for large-size stadiums required major advances in technology. Builders needed new materials to construct roofs that were at once wide enough to bridge a tremendous inner space and strong enough to withstand the resulting stress. Concrete had been stretched to its limits. European builders began experimenting in the 19th century with lighter structures made of a combination of glass and iron or of concrete reinforced with steel.

Public space for enormous exhibitions had suddenly come into demand. Proud of their industrial progress, nations of the world gathered in London to show off their products at the first world's fair in 1851. Star attraction was the exposition hall itself, the Crystal Palace. A shimmering vision of sheet-glass panels supported by a cast-iron and wood framework, it heralded a

new era of assembling large buildings out of mostly mass-produced parts. A singular virtue lay in its ease of construction.

Structures with glass vaulted roofs or curtain walls—from 19th-century shopping arcades such as Moscow's Trade Halls, which became the GUM department store, to Canada's modern West Edmonton Mall and Paris's Pompidou Center—are functional descendants of the Crystal Palace. Visitors stroll in spacious surroundings, undisturbed by weather or traffic.

Both exposition halls and sports arenas need a maximum amount of open space. An exhibition requires a large area

PRECEDING PAGES: Against the backdrop of a newer Rome, the ruins of the Temples of Saturn and Vespasian rise from the ancient Roman Forum. Heart of a city that was itself the heart of the Roman state, the Forum was the most historic of Rome's public spaces—the reputed spot where the city was founded by Romulus in the eighth century B.C. By the second century, the site—an open area used as a marketplace or for public assembly—was surrounded by shops, temples, courts, and monuments. Many of these buildings were constructed of a revolutionary new material—concrete.

Modern technology has freed recent generations of engineers and architects from the limitations of unreinforced concrete. Their structures now soar, creating huge public buildings undreamed of in Roman times.

SPACESHIP EARTH, EPCOT CENTER, FLORIDA

MARACANÃ STADIUM, RIO DE JANEIRO, BRAZIL

WEST EDMONTON MALL, ALBERTA, CANADA

to set off displays and accommodate the throngs of visitors. In an arena spectators want an unobstructed view of the playing field. As a result, most arenas and exposition halls lack interior bearing-walls, thus placing additional stress on their exterior walls and roofs.

Ways have evolved to control this stress. Steel frameworks, interior or exterior, support lightweight panels of glass, steel, or the plastic materials of the 20th century, producing some wondrously airy-looking structures, such as the plexiglass-roofed 1972 Olympic stadium in Munich. In another lightweight solution, roofs of air-inflated fabric reinforced by steel cables, such as those of the Pontiac Silverdome in Michigan and the Hoosier Dome in Indianapolis, span several acres.

Perhaps the techniques that built these pleasure domes will one day aid whole communities. R. Buckminster Fuller, inventor of the geodesic dome that inspired the Spaceship Earth sphere at Epcot Center, imagined a dome covering a portion of Manhattan—the ultimate air-conditioned environment. German architects Frei Otto and Ewald Bubner have designed a 71-acre, climate-controlled city for the Arctic, roofed by an air-pressurized membrane. Such technology could greatly improve conditions for people living in the world's harsher environments.

Sports Arenas

An ancient marvel in size and scale, the Colosseum owes its longevity to the strength of Roman concrete.

Among the largest and most imposing of all public places are those built for sports, and today, as in the days of ancient Greece and Rome, most sizable cities are home to at least one stadium or arena.

In the fourth century B.C. the Greeks erected a stadium at Olympia, site of the original Olympic Games. Primitive in design, the stadium took advantage of the topography. One side of the field nestles against a hill; heaped earth around the other sides forms huge, sloping embankments for spectators. Other Greek builders refined this design, cutting terraces into hillsides and lining them with stone seats.

In Roman hands the Greek stadium became the freestanding stone amphitheater, the largest of which was the Colosseum. These amphitheaters fell to ruin with the fortunes of the Roman Empire. Not until the late 19th century, with the renewal of interest in large-scale spectator sports, would such huge arenas again be built.

In the 20th century the evolution of football, ice hockey, and other sports has led to stadiums tailored to particular needs, although in recent decades the trend has shifted toward more versatile designs. Houston's 1965 steel-roofed Astrodome, with a seating capacity of 74,000, was among the first of these multiuse, roofed stadiums. It was succeeded by Seattle's Kingdome in reinforced and prestressed concrete, and then by the Louisiana Superdome in steel. Fabric-roofed structures have also appeared. Air pressure holds up some, such as Pontiac, Michigan's Silverdome; others, like the Munich Olympic Stadium, take a tentlike form.

THE LOUISIANA SUPERDOME, OPENED IN 1975.

The Colosseum

As enduring as the Eternal City itself, the Colosseum stands as a haunting reminder of Roman grandeur. Even so, two-thirds of the original structure has vanished, lost to lightning and earthquakes, vandals, souvenir hunters, and medieval construction crews who used the building as a quarry.

What remains owes its permanence to a design as rock-solid as the Colosseum's 40-foot-deep foundation. Construction took ten years and got under way in A.D. 70 on orders from the emperor Vespasian. He envisioned nothing less than the largest structure of its kind—615 feet long, 510 feet wide, 159 feet high, and more than a third of a mile in circumference.

Ground was not so much broken as drained, since the emperor had chosen an artificial lake as the site for his new amphitheater. Concrete soon replaced water, and powerful travertine piers, set on the concrete foundation, were linked radially by arches, walls, and vaults—mostly of brick-faced concrete—to make five concentric rings. Upon this structure more arches, walls, and vaults were erected to support three levels of sloping stone seats. Wooden bleachers stood at the fourth level, backed by the unwindowed wall carried upon the exterior arcades.

The builders of the Colosseum were

Sunlight plays across the interior of the Colosseum and spills into the amphitheater's underground chambers and passageways. In Roman times, they would have been concealed by removable wooden flooring.

The Colosseum

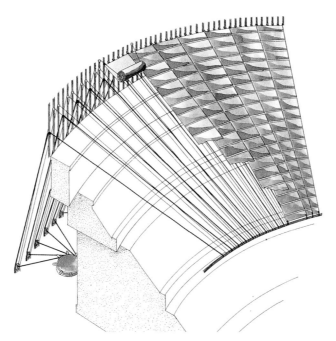

A net of ropes—controlled by winches manned by teams of marines—supported the *velarium*, a movable canvas awning that sheltered spectators from the hot Mediterranean sun. A beating drum timed its emplacement.

games and wild beast hunts—often cruel and brutal. Boxers, horsemen, and archers might perform, but it was the blood sports that drew the cheers. For most events, the arena was spread with sand, the better to absorb spilled blood, although on occasion it was flooded with water to set the stage for a mock naval battle.

The mechanical devices used to flood the arena were housed below the removable wooden arena floor. Here, too, were the hoists for the sets and scenery, as well as an extensive and ingenious system of passageways, cages, and cells, which held beasts and gladiators.

as attentive to aesthetics as to structural principles, for they created an edifice whose structural ingenuity was matched by its elegance, both inside and out. Smoothed travertine formed the facade of the building, the lower three stories of which were articulated by 80 arches framed by column-like forms.

On the fourth floor, the exterior wall was studded with stone brackets, divided into sections by tall Corinthian pilasters, and capped by a cornice. Holes in the brackets once acted as sockets for the wooden poles that supported a huge awning that could be unrolled.

In all, as many as 45,000 to 55,000 spectators could crowd into the Colosseum for a full day's worth of free entertainment, played out in the central arena. A parade of chariots might open the spectacle, which would include gladiatorial

An aerial view of the Colosseum shows its striking similarity to many modern stadiums and arenas. Concentric tiers of seats could hold as many as 55,000 spectators.

The interior of the Colosseum was a honeycomb of vaulted corridors and stairways, built mainly of brick-faced concrete. To facilitate crowd control, ticket numbers directed spectators through some 80 entrances.

The subterranean level clearly demonstrates Roman engineering skill. From here, animals in cages were hoisted to the upper level. Once the cages were opened, the beasts could escape—but only up a ramp and through a trapdoor into the arena (red arrows). Nets on top of the wall surrounding the arena provided extra protection from rampaging animals.

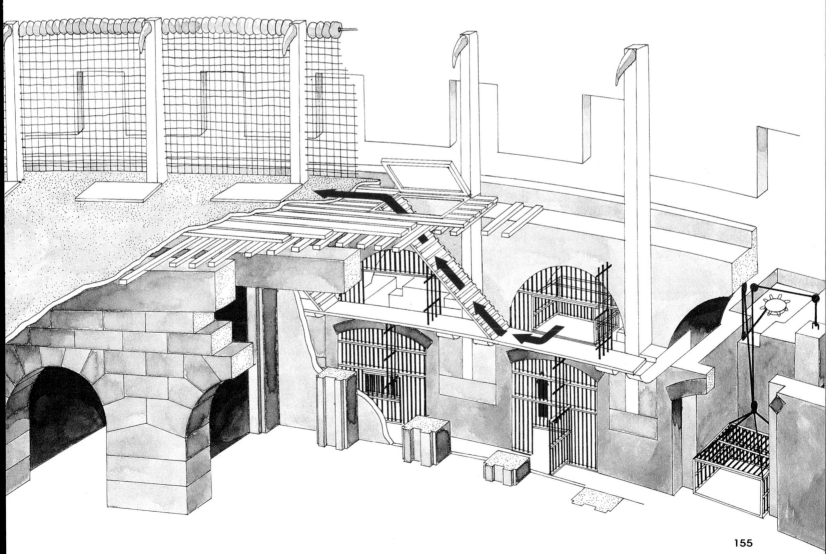

The Louisiana Superdome

Superdome workers unroll artificial turf, nick-named "Mardi Grass," in preparation for a baseball game. Movable seats along the lower concourse have been fully retracted to make full use of the playing area. Raised well out of range of the loftiest fly ball, a 75-ton gondola (center, top) houses giant TV screens, lights, and other apparatus.

Rising 27 stories above the streets of New Orleans and covering 10 acres, the Louisiana Superdome is the largest of all enclosed stadiums, "large enough," in the words of project director Nathaniel Curtis, "to house the most spectacular extravaganza and small enough to accommodate a poetry reading."

A system of movable grandstands gives the Superdome its versatility, allowing the stadium to expand from a compara-tively compact arena seating as few as 20,000 spectators, to a football stadium with room for 84,000 fans, to a convention or concert hall that can hold more than 103,000 people.

The building posed a challenge to its architects, the New Orleans office of Curtis and Davis, and its engineers in the firm of Sverdrup & Parcel, Inc. The steel dome was the largest ever attempted, and the sheer weight of the edifice, with its 169,000 cubic yards of concrete and 20,000 tons of steel, would generate massive compression.

To cope with the compressive forces, the design called for some 2,100 pre-stressed concrete pilings to be driven 165 feet to bedrock. Five thousand additional pilings would form the underpinning for the stadium's underground garage and for the concrete slab beneath the playing field.

As for the dome, a skeleton of steel ribs provided its backbone. These radial

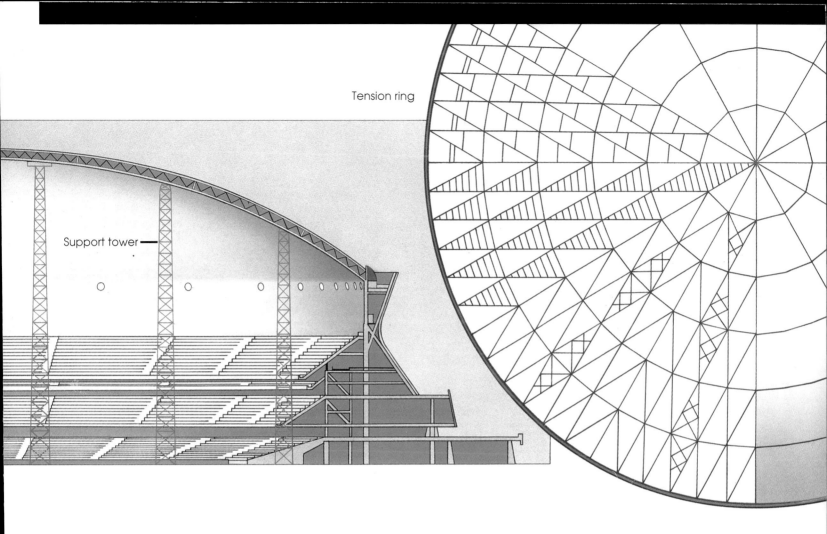

Tension ring

Support tower

ribs were linked by a series of concentric struts and further strengthened by cross-bracing to create a framework of hundreds of small lattice trusses.

To facilitate the roof's construction, the builders—the Alabama-based Blount Brothers Corporation and Huber, Hunt, and Nicholls of Indianapolis—set up temporary supports. Once complete, the roof was jacked down as a single unit onto its

permanent support—a huge steel tension ring 680 feet in diameter.

The tension ring is one of the most important structural components of the Superdome; without it, the dome would collapse. A circular truss made of $1\frac{1}{2}$-inch-thick steel, the ring was prefabricated in 24 sections, each of which had to be lifted into position, 169 feet up, before being welded to adjoining sections. The welding

A diagram of the roof's framework shows the tension ring and the lattice trusses created by intersecting radiating ribs, concentric struts, and cross braces. In the cross section, the roof's temporary support towers are visible.

A sampling of seating configurations reflects the Superdome's versatility. Seat sections can be moved along tracks recessed into the floor.

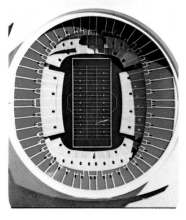

Football

Baseball

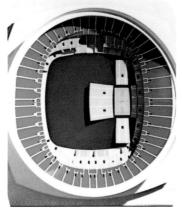

Arena

Convention

The Louisiana Superdome

The tension ring, which channels the dome's outward thrust onto vertical supports, encircles the superstructure. Behind, a gutter tub flares outward to catch rainwater—part of an ingenious system that regulates water outflow to prevent overtaxing the city's sewers.

A forest of 37 temporary support towers (opposite) crowds the Superdome's interior during roof construction. Each tower stood on its own pile foundation, and each had a hydraulic jack mounted on its top. Once the roof was complete, the jacks lowered the entire structure—all 5,000 tons of it—onto the tension ring. The towers were then removed. At the center of the photograph is the weblike "crown block," from which the dome's steel ribs radiate.

took place in the controlled atmosphere of a tent-house that was moved from joint to joint as work progressed. Finally, each weld was X-rayed to ensure its integrity.

None of this, however, was enough to convince the roofing subcontractor that the dome's design wasn't fatally flawed. Indeed, given the immensity of the dome and the added stress imposed by normal variations in temperature and wind pressure, he had reason for concern; hurricane threats only heightened his anxiety.

The designers, however, had already taken such factors into account and had designed a roof with an unjointed skin that would "breathe," rising and falling as much as three inches in response to fluctuations in the temperature and wind pressure. In addition, the pinned connections linking the vertical columns to the tension ring allowed up to eight inches of movement.

With the original design vindicated, work on the roof resumed, albeit with a

It is hardly surprising that the world's largest soccer stadium is located in Brazil, a country whose enthusiasm for what the locals call *futebol* borders on hysteria.

Although its cantilevered roof is only slightly more than 75 feet high, Rio de Janeiro's Maracanã Stadium can seat 150,000 people, with standing room for another 50,000. Visibility from the grandstands was scientifically calculated; there are no vertical piers to impede views, and no seat is more than 375 feet from the field. For safety and convenience, the ramps are wide enough to empty the stadium in 15 minutes, and, to protect players and officials from over-zealous spectators, the designers encircled the playing field with a five-foot-deep dry moat.

Maracanã was built of reinforced concrete in just two years by a labor force of more than 10,000, working day and night. It was completed in time to host the 1950 World Cup soccer championship. Since then, a huge sports complex has grown up around the stadium, with world-class facilities, including a domed gymnasium seating 20,000 people.

new subcontractor. Panels of 18-gauge sheet steel were hoisted into place, covered by an inch-thick layer of polyurethane foam, and topped by a layer of waterproof hypalon plastic. Soon thereafter, the temporary supports were carefully removed, allowing the tension ring to shoulder the entire weight of the roof. As it did so, the Superdome settled another $3\frac{1}{2}$ inches under the weight.

It took another two years to complete construction. By August 1975, with the last seats bolted down and thousands of square yards of artificial turf in place, the Superdome was ready to host its first event, a football game between the hometown New Orleans Saints and the Houston Oilers.

A designer of the Munich stadium's spectacular roof, German architect Frei Otto trained as a stonemason. He turned to architecture after an interest in model airplanes led him to study the behavior of membranes stretched on lightweight frames. By the mid-1950s, still in his 30s, Otto was experimenting with tentlike, cable-supported structures covered by cotton canvas roofs. A cable-net pavilion at the 1964 Swiss National Exhibition at Lausanne and another at Montreal's Expo '67 set the stage for his work on the stadium. If there is an irony in its enduring distinctiveness, it lies in Otto's reluctance "to fill the earth's surface with lasting buildings."

Anchored by masts outside the stadium, the roof has no supports to obstruct the view. The plexiglass allows light to filter through, while still protecting spectators in the west stands.

Gleaming gossamerlike in the bright sun, the canopy roof (opposite) seems to flow in waves from the stadium to the nearby gymnasium and swimming center, which it also covers.

The Munich Olympic Stadium

The year was 1972, the place Munich, and what was an abandoned World War II airfield had been reworked into a landscape of hill and hollow, meadow and lake as a setting for the Olympic Games.

The centerpiece of the Olympic Park was, and is, its striking stadium, dominated by a dramatic, tentlike plexiglass roof.

Ordinarily, a roof might be expected to be in compression, its weight borne by trusses or a lattice of support beams, or, like a dome, to generate outward thrust. But the roof of the Munich stadium is actually in tension, its weight carried by a prestressed cable net curved in two directions to stabilize it in the wind. The net is in turn suspended from hollow steel masts anchored in concrete and bolstered by steel guy cables. Positioned outside the stadium, the masts do not obstruct views of the field.

Thousands of four-inch-long bolts connect the cable net and the plexiglass canopy. They are attached directly to the individual plates of plexiglass, rather than to the aluminum frames in which the plates sit. As a result, the weight of the plexiglass is transferred to the net, not to the aluminum frames, lessening the chance that any of the plexiglass plates might be wrenched from their moorings.

The choice of plexiglass as a roofing material was dictated by, of all things, the requirements of television, marking one of the first times that television had a direct impact on a stadium's design. Unlike other materials, plexiglass would not cast heavy shadows, something early 1970s television technology could not contend with.

Rocking unsteadily in a safety net, two workmen repair the network of cables supporting the plexiglass roof.

Exposition Halls

When completed in 1982, Epcot Center's Spaceship Earth was on the cutting edge of design and technology.

Few public places have a greater need to enclose more space with the least amount of material than those built to house expositions. Large exposition halls are a relatively modern phenomenon, the product of lighter and stronger building materials that can span great spaces without collapsing. In fact, it wasn't until 1851, with the building of the Crystal Palace in London, that the exposition hall truly came of age and, with it, the modern era of pre-fabricated, modular construction.

Over the ensuing decades, "crystal-palace construction" became the inspiration for other temporary exposition halls, as well as for permanent arcades, galleries, and other building types worldwide. Later in the 19th century, the introduction of space-frame construction, in which stresses are equally distributed within a three-dimensional structure composed of inter-connected members, marked another advance in the evolution of the exposition hall. The form achieved a kind of perfection in 1982 with Epcot Center's Spaceship Earth in Florida, the first monumental geodesic sphere.

THE CRYSTAL PALACE, LONDON, 1851.

The Crystal Palace

It was the Great Exhibition of the Works of Industry of All Nations, the first world's fair, and when it opened in 1851 more than eight miles of tables displayed such wonders as false teeth, artificial legs, and chewing tobacco. One contraption, lampooned by a London newspaper as "a cross between an Astley Chariot, a wheelbarrow, and a flying machine," was the McCormick reaper.

But for many of the six million people who visited the Great Exhibition in London that year, the most intriguing object

Prototype for today's exposition halls and precursor of modern construction methods, the Crystal Palace measured 1,848 by 408 feet and covered 19 acres. It was, in effect, a colossal display case exhibiting the marvels of 19th-century industry and commerce.

Iron columns and glass-sheathed roof trusses created a vast interior space awash in daylight. The transept, reaching 108 feet high, easily held a fountain and several live elm trees.

Sir Joseph Paxton (1801-1865) was a 49-year-old horticulturist when he submitted an 11th-hour bid to erect a structure big and bold enough to house the Great Exhibition of 1851. His Crystal Palace challenged the technology of his time and profoundly affected subsequent building, and it earned him a knighthood and, later, a seat in Parliament. He went on to design conservatories, houses, lodges, and even a castle.

displayed was not inside the building but the building itself, the "Crystal Palace," as it came to be known. And this stunning work had been conceived not by an architect but by a horticulturist named Joseph Paxton, a farmer's son with little formal education. Its design had taken just eight days and the construction of its entire frame a mere 17 weeks.

Still considered one of history's most influential buildings for the construction method it introduced, the Crystal Palace looked like an overgrown greenhouse, its long, stepped, rectangular main hall bisected by a high, vaulted transept and topped by a pleated, ridge-and-furrow roof. Hollow iron columns doubled as downspouts, carrying off any rainwater that collected in the fluted rafters that

formed the furrows of the roof. Diagonal wrought-iron rods, crossed to form X-shaped portal braces, linked each horizontal beam to pairs of vertical columns. Together with the roof's trussed girders, these portal braces gave the exterior walls the rigidity they needed to compensate for the absence of interior walls.

In all, the building held some 4,500 tons of cast and wrought iron, 600,000 cubic feet of wood (used in the gutters, mullions, and interior arches), and 900,000 square feet of sheet glass; it covered

19 acres of London's Hyde Park. Yet despite its size, the Crystal Palace had a modular design and prefabricated components that allowed it to come down as easily as it had gone up. In fact, the design of the Crystal Palace incorporated just two different story heights, three different widths, and seven different iron components. Hence, many of the components could be cast in standard lengths at foundries elsewhere in England, brought by rail to London, and assembled on-site.

The Crystal Palace's economy of

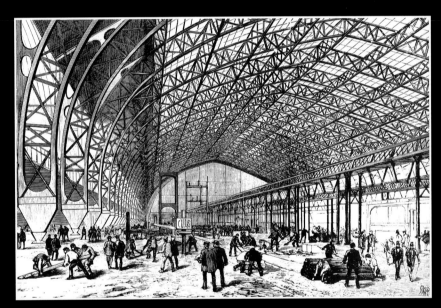

Designed to celebrate French industrial prowess, the 1889 Paris Exhibition also marked the centenary of the French Revolution. The Gallery of Machines, on the Champs de Mars opposite the Eiffel Tower, was itself an engineering triumph. Framed in the new harder and stronger material—steel—instead of iron like the Crystal Palace, the Gallery's glass panels were fixed to its exterior, shaping a vast inner, seemingly limitless, space. Twenty pairs of hinged girders formed arches spanning 380 feet and soaring to 140 feet at the apex. The pin supports at the arches' tops and beneath their bearing-plates allowed the building to flex if its metal expanded or contracted, or if the ground supports shifted. The strikingly innovative building was demolished in 1910.

Building the Crystal Palace

Simple in design, the Crystal Palace was built in just 39 weeks by teams of workers laboring simultaneously throughout the building. Although much of the work was mechanized, horsepower (below) helped hoist roof trusses.

Custom-built carts (upper) rolled along the roof's gutters, helping install some 300,000 panes of glass.

Workmen lock a column to a roof truss (lower), using cast-iron connecting pieces held in place by wooden or wrought-iron "keys."

design also streamlined its construction. Columns and girders, for example, were attached by cast-iron connecting pieces in a simple post-and-lintel configuration that allowed a team of workers to raise a self-supporting 24-foot module of three columns and two girders in just 16 minutes. Glazing the roof was equally simple, thanks to six "traveling stages" manned by crews of glaziers capable of installing as many as 19,000 panes of glass per week.

The end of the Great Exhibition late in 1851 marked the end of the need for the Crystal Palace, and true to Paxton's design, the structure was dismantled as readily as it was erected, then just as easily rebuilt on a new site on the outskirts of London. There, an even bigger and more elaborate Crystal Palace enjoyed a second life as a combination museum, concert hall, and amusement park, until it was destroyed by fire in 1936.

Encased in nearly one million square feet of glass, the Crystal Palace heralded today's glass-curtain buildings. Glass was also critical to the modular design, with the combined lengths of two glass panels partly determining the length of roof and gallery girders—and thus the entire building's dimensions.

Arcades

The glass umbrella of Milan's Galleria Vittorio Emanuele II (opposite) remains, as it has for more than a century, *il cuore della città*— "the heart of the city."

Opened in 1981, the glass-roofed "Europa Boulevard" of the West Edmonton Mall in Alberta, Canada, echoes in style and structure the shopping arcades of the 19th century.

Joseph Paxton's glass palace inspired dreams of more elaborate creations. He himself imagined a "Great Victorian Way," a glass-enclosed compound that would encircle London. Such ideas brought new life to a form that both grew up and died out in the 19th century—the arcade. The first vaulted glass-roof arcade was built about 1830 for the Galerie d'Orléans in Paris.

Arcades evolved from a need for protected public spaces that could handle the flow of crowds; glass had the advantage of at once providing light and shelter. Arcades served a number of social purposes—department stores, markets—often becoming centers of public life. They also served as models for other building types, such as the means of access in prisons, the railroad station, and the bath house.

One of the most influential arcades was the monumental Galleria Vittorio Emanuele II in Milan. Dedicated in 1867, it was designed as a covered promenade between the city's cathedral and opera house. Its massive dome, supported by arch beams, rises some 130 feet. Circumferential supports hold ribbing that secures overlapping panes of glass, through which light showers on a quarter-acre rotunda.

As the 19th century progressed, glass-enclosed arcades expanded in dimension and importance. The largest was Moscow's 1893 Upper Trade Halls (GUM), an 820-foot-long glass barrel vault with a very light roof—and multileveled accesses to more than 1,000 interior units. For the first time, an arcade was its own system of space, not a connection to something. It pointed the way to the American shopping center, a 1950s creation that has grown immense in Minnesota's 4.2-million-square-foot Mall of America, opened in 1992.

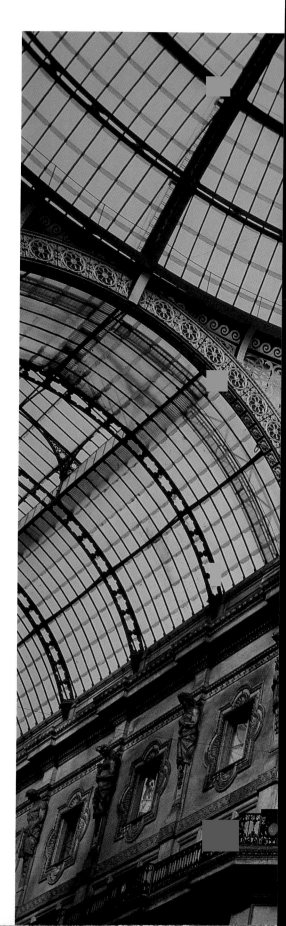

The Pompidou Center

The late French President Georges Pompidou once said all too prophetically of the cultural center that now bears his name, *"Ça va faire crier*—It's going to make for a lot of shouting."

Indeed, an outcry greeted the 1977 opening of the Pompidou Center in Paris. Critics likened the structure to a distillery or a refinery and denounced its architects,

the Pompidou Center was also built largely of prefabricated parts. The exoskeleton—a lattice of structural supports together with mechanical services—wraps around the glass walls. In fact, just about every structural element normally found inside a building is on the outside, leaving six floors of unobstructed space that can be tailored to fit the center's ever-changing needs.

Visitors to the center (below, upper) view the Beaubourg district of Paris from one of the walkways cantilevered outside each floor. Moving caterpillar-style, elevators enclosed in glass tubes (lower) link the walkways and enliven the transparent facade.

Richard Rogers and Renzo Piano, as "the musical comedy team."

Others, however, recognized this two-block-long building for what it truly is: yet another incarnation of the Crystal Palace, a glass box—suspended, in this case, from a visible steel exoskeleton—in which art of various kinds could be both displayed and created.

Avant-garde like the Crystal Palace,

On the Pompidou Center's east side (opposite), color-coded water pipes, electrical conduits, and air-conditioning ducts, along with steel supports, obscure the glass walls.

171

Spectacular variation on the modern three-dimensional space frame, Spaceship Earth consists of two interconnected nested spheres. The outer one's nearly 1,000 aluminum panels reflect an ever-changing pattern of color and light. The inner sphere is made of 1,450 steel beams covered with waterproof sheeting.

Spaceship Earth

eS een from a distance, it resembles the huge golf ball it has often been likened to, all teed up on its six steel legs and surrounded by the pavilions of Future World at Walt Disney World's Epcot Center.

Up close, however, Spaceship Earth is revealed as a complete geodesic sphere, the world's first, as well as its largest, with a volume of more than two million cubic feet.

The virtue of all such structures is their comparative lightness—their capacity to enclose large spaces using far less material than standard frames. Still, Spaceship Earth is massive—18 stories high and weighing 16 million pounds. It depends on an ingenious double-layered frame for stability. Triangular aluminum panels form a faceted outer shell over an inner framework of steel. The frames are interconnected, bolted together by steel struts and ties so that the load is distributed evenly.

The world's first geodesic sphere, Spaceship Earth dominates the horizon at Epcot Center, near Orlando, Florida. There, under construction in 1981 (below), triangular steel bracing, similar to that of Spaceship Earth's, prepares supports for the Land pavilion.

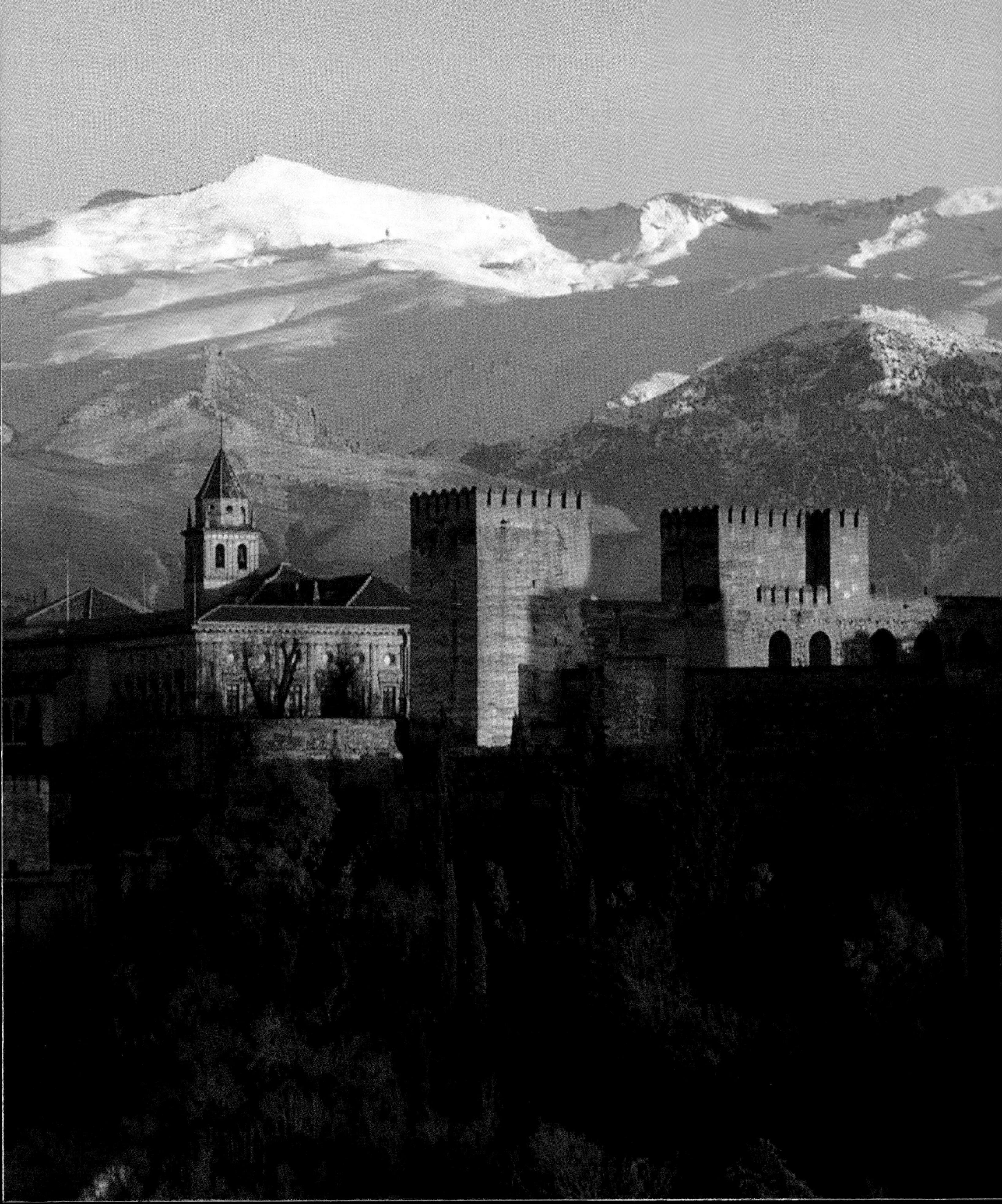

ON LAND
FROM WATER

People have built defensive walls since Neolithic times. As early as 8000 B.C., the city of Jericho was protected by a barrier 13 feet high and 10 feet thick. Babylon is said to have been surrounded in biblical times by 11 miles of defensive walls. Military strategy in classical Greece often demanded walls and counter-walls. Victory went not only to the warrior but also to the stonemason. In China, the Great Wall stands today as a testament to the tenacity of its builders.

In some civilizations, walls not only enclosed a city but defined it as well. The Chinese, for instance, used the same word—*cheng*—to describe the two.

Walled strongholds served both as defenses and as symbols of power and centers of control over urban populations. Openings in a wall were guarded as carefully as spillways in an irrigation system. Fortified gates discouraged attacking armies and encouraged respect. In Europe and elsewhere, walls endure as prominent features of cities such as York, Carcassonne, and Istanbul.

In 11th-century England, early castles and their defending walls were little more than high mounds topped with wooden towers and surrounded by a ditch with a series of barricades. Changes in castle construction came slowly, often in response to advances in weaponry. Wooden barricades gave way to stone walls. Towers with square corners that could be easily broken by siege equipment took on a variety of new shapes. Emphasis shifted away from tower defenses to the construction of virtually impregnable walls.

By the 14th century, the concentric

THE OOSTERSCHELDE SEA BARRIER, THE NETHERLANDS

THE THAMES BARRIER, ENGLAND

PRECEDING PAGES: Nestled against the snow-capped peaks of the Sierra Nevada in southern Spain, the Alhambra recalls the era of foreign conquest. On the site of a fort built in the ninth century during peasant uprisings against the Moors—North African Muslims who had conquered Spain a century earlier—the Alhambra commands a 35-acre plateau above the city of Granada. After Granada became the Moorish capital in 1238, Muhammad I fortified the Alhambra with thick ramparts and 22 towers. Inside, he and his successors created an Islamic version of paradise filled with ornamental fountains, patios, and gardens. No invader ever stormed the Alhambra's walls. Ferdinand and Isabella of Castile considered it, but they were preempted by the peaceful surrender of the last Moorish ruler in 1492.

castle—a circuit of walls and towers surrounded by another, lower circuit of battered walls with towers of its own— proved outstandingly successful against invaders. Only powerful siege guns, developed in the late 14th century, could force the castle's occupants into submission.

From earliest times as well, human settlement has been drawn to riverbanks and seacoasts. Proximity to water is crucial to both agriculture and commerce, but it also exposes large numbers of people to the risk of floods.

Some flooding, of course, brings

SACSAHUAMAN, CUZCO, PERU

benefits; for example, the fertile silt laid down each year by the Nile supported the great civilization of ancient Egypt. Just as often, though, flooding brings widespread devastation: Inundations caused by seasonal melts, severe storms, and flood tides have killed thousands of people and wreaked havoc on the land along coasts and rivers worldwide.

Ancient measures to protect areas against flooding include the building of earthen levees, dams, and dikes to hold back rivers and seas. Levees and dams protected cropland along the Nile as long as 5,000 years ago; linked with canals, they provided a sophisticated irrigation system.

By A.D. 1000 the Dutch added dikes to their network of levees and *terpen,* mounds of clay and sod that raised settlements above flood level. The dikes served as paths connecting terpen; where they crossed rivers, a simple sluice gate opened and closed with ebb and flood. Dutch engineers also provided their sea defenses with dikes, incorporating knowledge of waves, tides, and seasonal changes.

Modern flood control uses many of the older techniques but on a larger scale and with an ever increasing technological complexity. But even state-of-the-art projects such as the Dutch Deltaplan and the Thames Barrier may ultimately answer to their all-powerful enemy, the sea.

THE GREAT WALL, CHINA

On Land

Ramparts of the Great Wall once defended China against invaders from the north.

A rchaeology tells us that early civilizations commonly built walls. Made of wood or mud brick, and later of stone or fired brick, high defensive barriers became higher. They protected homesteads, towns, and even entire city-states.

Well into medieval times, walls were the key to defense against enemies on land, whether marauding neighbors or invading foreign armies.

Fortress construction changed over time. Earlier square or rectangular construction employed thick, high walls and few openings. Thick walls were effective walls. Engineers added bulk economically by building shells of stone or masonry and filling them with rubble. Walls within walls were even better; if the enemy broke through the outer defense, the defenders could rally from the inner enclosure.

Special attention was given to the keep, used as both residence and refuge.

Gradually, dependence on rectangular stone keeps—for passive defense—yielded to emphasis on circular towers of lighter construction. These allowed a wider field of vision and thus a more active defense.

The advent of gunpowder and other advances in warfare eventually rendered immense walled fortifications obsolete. But walls would take on a new purpose in military forts engineered to give superior advantage to the placement of cannon.

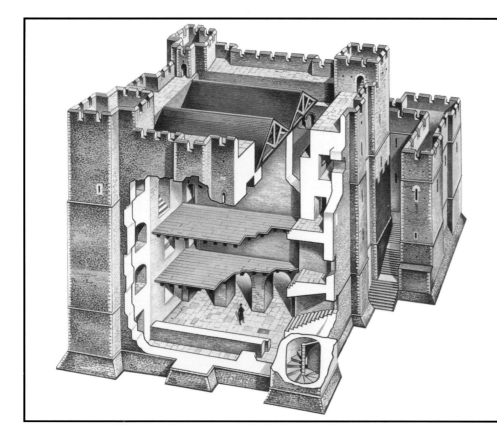

Atop Dover's white chalk cliffs, an imposing keep represented both an end and a beginning in castle construction. Built in the 1180s by Maurice the Engineer for Henry II, Dover Keep measured 98 feet long, 96 feet wide, and 95 feet high. These dimensions, and buttressed walls tapering from 21 to 17 feet thick, made it the largest and strongest of the Norman-style square keeps in England. Military and domestic functions were divided among three floors connected by spiral stairs. Concentric curtain walls with regularly spaced mural towers heralded castle design of the future. Another element—tower-flanked gates—dated back to Roman times.

Krak des Chevaliers

An aerial reconstruction from the northeast reveals a fortress designed to withstand long sieges. The arched main gate through the outer wall is in the foreground. The inner wall was built on higher ground to support the first line of defense. The keep—the three tall towers overlooking the outer moat—faces south, the most vulnerable side. A reservoir provided water; the windmill on the outer wall ground corn.

Massive walls of the Krak des Chevaliers (right) command a mountain spur more than 2,000 feet above a north Syrian plain.

Visiting the Krak des Chevaliers in 1909, T. E. Lawrence pronounced it "perhaps the best preserved and most wholly admirable castle in the world." Of the dense network of Crusader castles positioned throughout the Holy Land in medieval times, only the Krak has survived intact—a testament to its ingenious construction.

The Krak des Chevaliers (the name is an Arabic-French hybrid meaning "Fortress of the Knights") and the other castles were built to defend Crusader conquests from the Muslims. They were garrison forts, strategically located on prominent ground and spaced about a day's ride apart.

The Krak stands on a steep rise in northern Syria, on the site of a Kurdish fortress that came under Crusader control in 1099. In 1142, the fortress was given to the Knights of St. John, or the Hospitalers, a religious military order. They rebuilt it into the foremost Crusader stronghold, capable of resisting siege for months at a time.

The Krak's strength lay in its siting and design. Inner and outer curtain walls describe a concentric plan. The walls, interspersed with towers, are made of masonry blocks 15 inches high and more than 30 inches long and have a rubble-and-mortar core. On two sides huge taluses—slopes of soil and rock debris—protected the inner walls from earthquakes and from attempts to scale or mine them.

Enemies breaching the main gate met a labyrinthine entrance passage that delayed their progress and left them vulnerable to overhead fire. Only once in the Krak's more than 200 years of Crusader service did the Saracens reach the inner wall. Then, after a siege and weeks of heavy bombardment, the sultan Baybars took the fortress, in 1271.

Caernarvon Castle

P lagued by Welsh resistance, Edward I of England spent much of his 35-year reign building 17 fortresses to establish his presence in Wales. Many stood along the northwest coast, where they could be supplied and reinforced from the sea.

Grandest of these 13th-century castles, Caernarvon commands a site at the mouth of the River Seiont overlooking the Menai Strait. From its beginnings in 1283, Edward envisioned Caernarvon as a vice-regal center and the seat of the Prince of Wales; he bestowed the title on a son born there in 1284.

Edward was a hands-on royal contractor. He tried to be present at building starts and hired many of the artisans himself. For Caernarvon's design he relied on James of St. George, Master of the King's Works in Wales. A siege engineer as well as an architect, James understood defense from both sides.

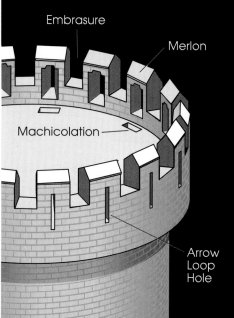

Embrasure

Merlon

Machicolation

Arrow Loop Hole

At a distance, medieval castles present immense, forbidding facades to the outside world. Close up, they reveal many innovations incorporated to give every possible advantage to the defenders. Several examples appear in this section of a castle tower rampart. Round towers let defenders assault attackers from all positions. Merlons, the solid sections of the crenelated wall, had arrow loops with walls splayed toward the inside to afford archers as little exposure as possible while allowing a wide angle of vision. Missiles also could be released from the open sections, called embrasures. Cauldrons of boiling oil or molten lead, as well as missiles, could be discharged through machicolations, openings in the overhanging floor of the rampart walk.

Caernarvon was built on a site containing the ruins of a Roman fort and a Norman castle. Its concentric plan measured roughly 570 by 200 feet, with an inner wall and an irregularly angled outer wall. On the exposed southern side the outer wall was thicker and higher, supporting three tiers of stations for defending archers. Thirteen polygonal towers protected the outer wall. Though symmetrically placed, no two towers are alike.

Two towers formed the castle's main focus: an immense gatehouse facing north

into the walled town of Caernarvon. It opened into a 60-foot passage plugged with six portcullises—retractable iron gates—and five sets of double doors. All this was protected from above by numerous arrow loops and seven sets of machicolations, or overhanging defensive structures. In place of a central keep, the Eagle Tower rose 124 feet along the western wall. It could be shut off and supplied from outside through its own water gate.

Aspects of Caernarvon's design, such as the polygonal towers and bands of

colored stonework, can be traced to the walls of Constantinople. Whether Edward saw similar structures on Crusade or read about them is not known. Scottish distractions and strained finances prevented both Edward and his son from using Caernarvon as a royal seat.

Spotlights illumine Caernarvon's crenelated walls and polygonal towers. The castle epitomizes the fortress building of Edward I.

The courtyard provides a dramatic setting for the 1969 investiture of the Prince of Wales.

183

The Great Wall

R egarded by many experts as one of the most ambitious engineering projects ever undertaken, the Great Wall of China as we know it today was built by hundreds of thousands of laborers over the course of several centuries—at a cost in lives that will never be known. Now partially restored, the wall extends some 2,000 miles from the desert in Gansu Province to the sea at Shanhaiguan on the east coast. It takes advantage of the defensive features of the land, often by following a ridge and then doubling back on itself to command the high ground.

Qin Shihuang, the First Emperor, began building border defenses in the third

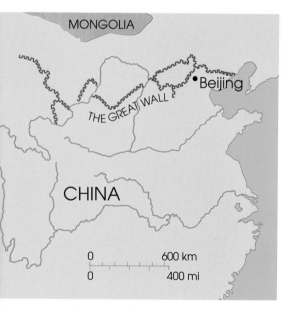

Built and rebuilt over several centuries, the Great Wall winds some 2,000 miles through northern China to the Yellow Sea.

The wall takes a tortuous path (right) along ridges in the mountains northeast of Beijing.

The Great Wall

The well-preserved section of the wall at Juyongguan (opposite) guards a strategic pass northwest of Beijing. The walls of the fort in the foreground bear inscriptions in the Chinese, Manchu, Mongolian, Arabic, and Tangut languages, evidence of international traffic.

century B.C. to protect his newly unified Chinese domain from northern invaders. He commandeered a huge civilian work force, as well as tens of thousands of soldiers. Sections of earlier walls built by local rulers were linked with new construction. The Qin wall lives on in popular legend, but almost no evidence remains to tell us of its location or the details of its structure.

The impressive wall of today was constructed almost entirely during the Ming Dynasty (1368–1644). It was built gradually, over time, in response to repeated threats from the north.

The earliest sections were earthen; later the builders began using stone. First they leveled the ground and laid courses of stone slabs for the foundation. Then they built the wall faces of stone and filled the space between with small stones, rubble, and earth. Layers of brick formed the top. Bricks and tiles that faced the foundations were prepared in on-site kilns, as was the lime used for mortar.

The Ming wall was battered: That is, its outer faces sloped inward as they rose. A typical section of the eastern wall stood 20 or 30 feet high and measured 25 feet thick at the base, 15 across the top.

The wall was punctuated at regular intervals by some 25,000 watchtowers, gates, fortresses, castles, guardhouses, and even temples and shrines. Beacon towers placed about 11 miles apart sent up signals of fire or smoke that warned other outposts of attack. Paths on the wall's crest permitted movement of troops.

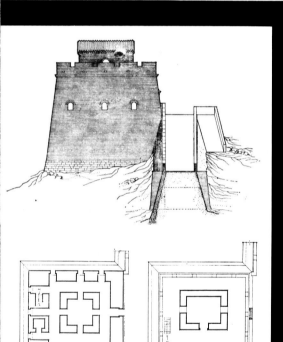

The Ming inscribed the names of engineers and supervisors on tablets along the wall but, like the Qin, neglected the nameless workers who died while building it.

Though a formidable barrier, the Great Wall never succeeded in stopping northern nomadic peoples or the Manchu invaders who became China's new rulers after they poured through the gates of Beijing in 1644. Although the wall has suffered extended periods of neglect, parts of it have been reconstructed many times, and the work continues today.

Throughout its approximately 2,000-mile course, the Great Wall incorporated a variety of defensive structures, many of which had proved effective in city walls. The drawing above shows a guardhouse erected near a gate in remote Shanxi Province during the Ming Dynasty, when almost all of the wall that exists today was built. A ramp on the inner side of the wall leads to the two-level structure (diagrams), which housed a small garrison. Capable of supporting cannon, the guardhouse, along with a nearby observation tower and signal tower, formed a defense and communications unit, one of many unifying security along the wall's entire length. Near its eastern terminus at Shanhaiguan, workers of the 1980s clear centuries-old foundations similar to those of the Shanxi structure. Building materials varied with the resources available at different locations. Here, tapering brick courses interfilled with rammed earth and underlain by a foundation of stone slabs give the wall strength and breadth. The Great Wall has always been a source of special pride for the Chinese. Repair of several sections continues today, using techniques little changed over the centuries.

Sacsahuaman

Jagged, terraced walls of Sacsahuaman (below), a 15th-century Inca stronghold, zigzag along a steep rise above Cuzco, in Peru.

Walls (opposite) incorporate stone blocks up to 25 feet high, all fitted by hand.

When 17th-century Spaniards encountered Sacsahuaman, the immense stone fortress that dominated the Inca city of Cuzco, in present-day Peru, they refused to accord it human origins. Black magic, the devil, and giants were proposed as more likely agents of construction than Pachacuti, the Inca ruler who started it about 1423.

Sacsahuaman was a bare hilltop north of Cuzco that each day caught the first rays of sunlight while the city remained in darkness. It was an ideal site for a fortress and refuge to which Cuzco's ruler and other inhabitants could retreat if city defenses were breached.

Everything about the construction of this Inca stronghold still amazes and much of it still puzzles experts who have studied it.

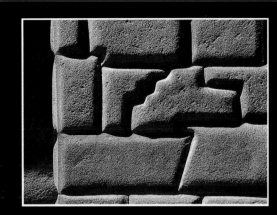

Inca builders took no shortcuts in their work. They often seemed to do things the hard way, perhaps, some archaeologists think, as a kind of bravado. Sacsahuaman's mortarless stonework attests to the high quality of Inca building, which made some Europeans believe that the Inca knew how to melt or soften rocks. With only stone and bronze tools, Inca masons shaped the face of each stone into a rounded bulge, tapering at the edges. With constant testing and cutting, the large, irregularly shaped stones were meticulously fitted together like pieces of a jigsaw puzzle. Sacsahuaman's walls, up to 60 feet high, still have joints so precise that a knife blade cannot fit between them.

Surviving structures include three terraced stone walls on the north side of the hill. The walls extend more than a third of a mile and zigzag like rows of sharp, menacing teeth. Originally they stood some 60 feet high; enormous stones, up to 25 feet tall and weighing up to 200 tons, formed their bases. Many smaller stones from the walls' higher courses are missing, pillaged by the Spanish for the building of post-Conquest Cuzco.

In Inca times, three brightly painted stone towers stood on the summit. The central Round Tower served as imperial palace; the other two housed the fortress garrison. Underground passages connected the towers and their courtyards.

There is evidence that Sacsahuaman benefited from a supply of pure water carried by stone conduit from a mountain reservoir. Inca genius obviously included hydraulic engineering; water was delivered to the Round Tower—some 120 feet above the surrounding terrain. Experts suggest a pressure-siphon system that allowed water to rise by means of its own pressure to a distribution tank. From the Round Tower, gravity would transport water to the rest of the complex.

To get Sacsahuaman started, Pachacuti commandeered 20,000 laborers from his empire. Some were designated as quarrymen, some as stone haulers, and others as stone fitters and wall builders.

Individual workmen came and went over the 60 years of construction.

Sacsahuaman's stonework is remarkable for any era. It is still not known how the large stone blocks were hauled to the building site, often over many miles, without benefit of wheeled transport. With only stone hammers and bronze crowbars, masons dressed and custom fitted each stone in place, creating tight, mortarless joints. So heavy yet flexible were the walls that they have survived earthquakes.

Sacsahuaman fell to the Spaniards and their scaling ladders in the 16th century. Soon after, the conquerors began the stone-by-stone dismantling of this marvel of Inca artistry and engineering.

From Water

Workers prepare a huge mat for placement on the seabed during the construction of the Dutch sea barrier.

The option offered to Noah as a defense against flooding was unique. Most inhabitants of flood-prone areas—now estimated at one-tenth of the world's population—have had to develop other means of protection against overflowing rivers and encroaching seas.

Over the millennia, flood-control engineers have pursued two basic strategies: building barriers such as dikes and levees or lowering water levels. The latter approach requires the storage of water in swamps or mountain reservoirs for controlled or gradual natural release.

These strategies hold worldwide, though the technologies used may vary.

Both Bangladesh and the Netherlands, for example, are low-lying countries imperiled by the sea. With few resources except manpower, Bangladesh relies chiefly on traditional technology, while the Netherlands draws on centuries of flood-control experience and the latest advances in hydraulic engineering as well.

BARRIERS OF CLAY-FILLED BAGS IN BANGLADESH.

The Guan Xian Flood Project

very spring, according to a 19th-century account, people in the western Chinese province of Sichuan would attend a ceremony dating back more than 2,000 years. Before dawn, a visiting dignitary would burn incense and offer prayers to the gods and to the memory of two third-century B.C. officials, Li Bing and his son, Li Erlang. Then he would leave the temple erected for the worship of Li Erlang and proceed to the bank of the Min River. There, the people would watch as a band of laborers pulled a bamboo cable to breach a temporary cofferdam, unleashing the waters of the Min for irrigation.

The diversion of the Min River forms the basis of the Guan Xian flood control project, begun about 250 B.C. The Min runs low in winter but is subject to heavy flooding during spring thaws and summer rains. Controlling it meant an end to droughts and floods, and transformed this area of China into one of its most productive.

The visionary behind the project, Li Bing, was administrator of the ancient province of Shu. His plan divided the Min into an inner channel for irrigation and an outer one to carry normal flow and runoff, as well as river traffic to the Yangtze.

Li Bing accomplished this using hand labor and local materials of wood and stone. His workers began by piling stones in the middle of the Min's natural channel, building a long embankment, or division-head, to divide it.

To direct the inner channel onto the Chengdu Plain, they had to cut a 90-foot-

wide canal through a nearby mountain. The water thus diverted eastward was channeled through a system of feeder canals, conduits, spillways, and lesser conduits to supply 2,000 square miles of farmland and serve five million people. Through a spillway, excess water was rerouted to the outer channel and on downriver.

After Li Bing died about 240 B.C., the project was completed by his son. Li Bing

An engineering project begun in the third century B.C. still diverts waters of the Min River in China's Sichuan Province. A man-made island (center) divides the Min into two channels, the inner (right) and outer feeder canals.

The illustration opposite shows the channel division and the canal called Neck of the Precious Bottle that cuts through a mountain to irrigate more than 1,400 square miles of farmland today. The spillway sends surplus water to the outer channel and on downriver.

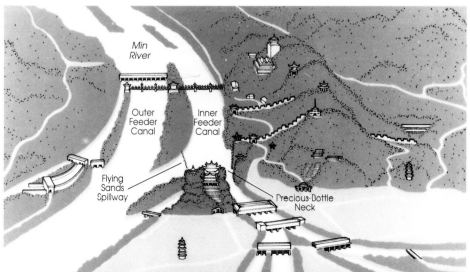

left a full set of instructions for its maintenance; his advice to "clear out the beds and keep the dykes and spillways low" has been heeded. In the past 2,000 years, Li Bing's scheme has been expanded but not greatly altered. Succeeding generations seem to have followed the rules of river control inscribed at Li Erlang's temple that read, in part: "Respect the ancient system And do not lightly modify it."

Min River

Outer Feeder Canal

Inner Feeder Canal

Flying Sands Spillway

Precious Bottle Neck

The Thames Barrier

The Thames Barrier spans the river eight miles downstream from London. Movable gates rise 60 feet against surge tides; during this early test, one remains open to shipping.

The Thames Barrier, a pier-and-gate structure in the river's channel eight miles downstream from London, stands ready to forestall disaster. With surge tides threatening and—over the long term—southeast England sinking and ocean levels rising, all factors point to inevitable flooding that would devastate a 45-square-mile floodplain where 1.7 million people live.

A 1953 tidal surge that inundated parts of London and killed 300 people downstream spurred construction of the barrier. Designed by the engineering firm of Rendel, Palmer & Tritton, the complex structure includes ten rotating steel gates spanning the 1,700-foot-wide channel at

Each of the steel-clad shells (right, upper) houses a giant rocker beam (lower) that rotates the gate-closing mechanism.

The barrier's gates operate on the same principle as a gas cock. As in the diagram, open gates allow water through; when closed, they shut off the flow.

Woolwich. Flush with the riverbed when open, the gates rotate 90 degrees to a height of 60 feet when closed, an operation that takes 20 minutes. The gates are separated by nine huge piers topped with steel-clad shells housing machinery that operates the gates.

In 1974, an Anglo-Dutch consortium—Costain Civil Engineering, Tarmac Construction, and Hollandsche Beton Maatschapij—began construction of the barrier in sections, which allowed river traffic to continue. Cofferdams created dry areas for building the six-story piers with their mazes of passageways, gate-support structures, and controls. The 10,000-ton concrete sills that support the gates were cast on-site, floated out into the channel, and lowered into place. A Liverpool company fabricated the shells of laminated wood clad with stainless steel and transported them to their Woolwich destination.

Costing more than half a billion dollars, the Thames Barrier project was completed in 1984. Already experts are looking ahead some 50 years, when rising tides could exceed the barrier's height.

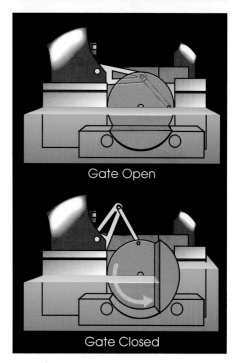

Gate Open

Gate Closed

The Dutch Sea Barrier

Giant mattresses fabricated to stabilize the seabed of the Oosterschelde estuary unroll like paper towels from the construction vessel *Cardium*. Preparing the seafloor was the first step in the erection of a two-mile-long open surge barrier, the final phase of a 30-year, 5-billion-dollar project to protect the southwestern coast of the Netherlands from the ravages of the North Sea.

The Dutch have always battled the encroaching sea. One-fourth of the Netherlands lies below sea level, and more than half is prone to twice-daily flooding by normal tides. Almost a thousand years ago the Dutch built their homes on artificial mounds above the sea's reach. Later, they used windmills to drain the land and constructed dikes and dams to hold back the tides. But the high waves of severe North Sea storms could easily breach these barriers.

A savage storm tide in January 1953 inundated the southwestern delta region, killing nearly 2,000 people and prompting the country to undertake the world's most ambitious and sophisticated flood protection project. Known as the Deltaplan, or Delta Project, it includes four large dams between the Westerschelde and the New Waterway. Its eight-year final phase, completed in 1986, involved the construction of a two-mile-long surge barrier in the Oosterschelde estuary.

In the original plan, the estuary was

Hydraulic mechanisms that lift each of the barrier's 62 steel gates rise from their sunken, 18,000-ton concrete piers (above). In the open position, the gates maintain normal tidal flow, allowing commercial marine life to flourish. Ingenious floodable construction docks (below)—built some 45 feet below sea level—permit each group of finished piers to be hoisted by gantry cranes on the vessel *Ostrea* and carried to final positions in the estuary.

to be closed completely, but conservationists persuaded the government to construct an open barrier instead. This movable barrier preserves the estuary as a commercial fishery and an important migratory-bird haven, but can be closed during storms to prevent flooding.

Before construction of the surge barrier could begin, many technical and logistical problems had to be overcome. The Dutch built several work harbors and yards for transferring and storing construction materials, as well as work sites on three artificial islands built on sandbars in the Oosterschelde itself. Two of the islands were then connected by a dam. This left the estuary permanently divided into three channels, each of which would receive a section of the surge barrier.

Central to the barrier design are 65

A three-layer sandwich of sand, fine gravel, and coarse gravel (above) reinforced with wire makes up each erosion-preventing mattress. The rugged fabric covers (below) are designed to stretch lengthwise but not crosswise. A worker marks each mattress with lines and section numbers that will act as guides for repairs after placement.

The Dutch Sea Barrier

New concepts in dam design and a fleet of specialized vessels made the sea barrier possible. Many elements provide a firm base for the movable gates that regulate tidal flow in the estuary. Stabilizing mattresses support sand-ballasted concrete piers; an armor layer of large rocks provides additional weight. Sills and beams define openings for the 400-ton gates, while a duct in the roadway beam houses hydraulic and electronic equipment.

gigantic concrete piers, each weighing 18,000 tons. These support the 400-ton steel gates, their hydraulic lifting mechanisms, and the massive sills and beams that stabilize the whole structure, as well as the roadway that caps the barrier. Each pier was under construction for about 18 months at an island work site. Starting a new pier every two weeks meant that dozens were under way simultaneously.

The hoisting and transport needs of the plan far surpassed the capabilities of any crane ever erected. Once completed, each pier had to be lifted, carried, and then installed in its precise position in the tidal basin. To accomplish this, engineers designed and built a U-shaped lifting

vessel, the *Ostrea,* equipped with tall gantry cranes and computers to target and guide placement. They also designed five additional vessels to perform other specialized tasks.

Before any part of the barrier could be put in place, the seabed had to be stabilized. This was done by means of giant vibrating needles extending from the vessel *Mytilus;* the needles compact the seafloor much as the soil in a flowerpot is

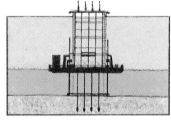

Mytilus

Cardium

Macoma, at left, and *Ostrea*

Trias

Taklift 4

compacted by repeated tamping. Further stabilization for the bases of the concrete piers is provided by 656-by-138-foot mattresses, filled with sand and gravel, that prevent seabed erosion.

Assembled on one of the construction islands and installed by the vessel *Cardium,* the fabric-covered mattresses update an older Dutch technique of using bound tree-branch mats to anchor earthen dikes. Smaller mattresses were laid on top of the larger ones in the exact position for each pier base.

Once installed, the piers received additional ballast in the form of sand pumped into their hollow caisson bases. *Taklift 4,* one of the world's largest floating cranes, hoisted into place the crossbeams, sills, concrete caps, gates, and roadway. Then pontoon vessels, such as the *Trias,* carefully heaped hundreds of tons of rock at the structure's base. This last step counteracts increased tidal pressure on the barrier caused by a three-fourths reduction in the estuary's discharge opening.

Building an open surge barrier that could handle tremendous tidal currents demanded ingenious solutions from hydraulic engineers. As one put it, "We had to invent everything from scratch."

Engineers frequently test the barrier gates, taking about an hour to open or close the system. Thirty-one operating strategies can respond to any kind of storm. In the spring of 1990, the North Sea administered a "final exam" by unleashing a storm of hazardous proportions. The surge barrier passed with flying colors and should protect the Netherlands for centuries to come.

Thirty-one gates span the two northern channels of the Oosterschelde barrier. For the time being, Dutch engineers have tamed—but not conquered—the North Sea.

The Feni River Closure

At low tide, Bangladeshi workers carry 100-pound bags of clay across the muddy bottom of the Feni River to waiting stockpiles. Later, during a frenetic seven-hour marathon, 15,000 laborers shifted the top layers of bags into the gaps alongside the stockpiles, creating the largest estuary dam in South Asia.

Like the Netherlands, tiny Bangladesh maintains a fragile coexistence with its benefactor and enemy, the sea. Bangladesh is a low, flat, densely populated country about the size of Wisconsin and composed almost entirely of the delta of the Ganges, Brahmaputra, and Meghna river systems. Rainy-season flooding regularly deposits a new layer of fertile soil on the hard-worked land. But seasonal cyclones can bring devastation and death as water from the Bay of Bengal surges across the delta, quickly flooding shallow rivers like the Feni.

Closing the mouth of the Feni River to protect against storm surges from the sea and to conserve river water for irrigation posed an immense challenge to this impoverished country. With only meager financial and technological resources, Bangladesh wanted a plan that would

take advantage of its few assets—abundant unskilled manpower among its 118 million inhabitants and natural resources suitable for low-tech solutions. To design the plan, they chose Hans van Duivendijk, a Dutch engineer with experience in his country's traditional flood-control methods.

The mouth of the Feni River is almost a mile wide. As in the Dutch Deltaplan, the first step was to protect the riverbed from erosion. Huge mats were laid and then ballasted with boulders trucked and barged from the north. Clay-filled bags and more boulders were dumped in gullies to create a level sill. To construct the dam, more bags were stockpiled across the channel until closure.

On a day chosen to coincide with the dry season's lowest tides, some 15,000 workers labored in a seven-hour frenzy to close the gaps. Afterward, dump trucks and earthmovers added clay to increase the dam's height to 30 feet. Finally, the dam was faced with concrete and brick and paved with a one-lane asphalt road.

Soon after the 1985 closure, a cyclone swept through the Bay of Bengal—but the Feni River Dam held. Its success holds promise for other low areas of Bangladesh and elsewhere in South Asia.

To stabilize the riverbed, workers laid mats (below) positioned by frameworks of bamboo and reed and weighted down with boulders.

The dam (below, lower) closes the Feni channel, then doglegs across a sluice-controlled bypass to connect with coastal embankments.

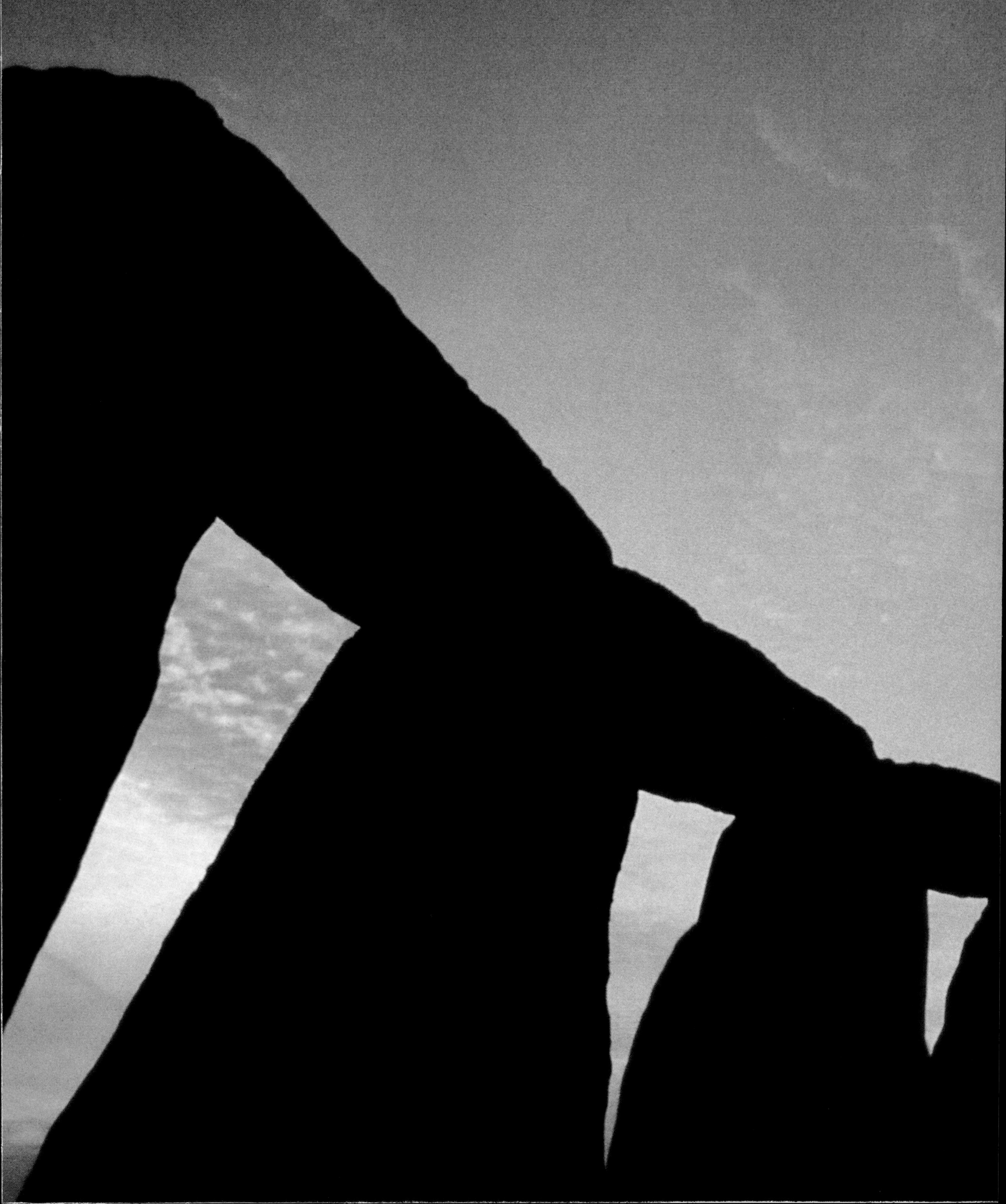

PYRAMIDS
TEMPLES
DOMES
GOTHIC CATHEDRALS

THE PYRAMIDS AT GIZA, EGYPT

Humans have long made structures that stretch the fabric of belief into palpable form, whether as places of worship, strongholds of god-kings, tombs, or observatories to probe the cosmos.

Historically, the quest for the eternal accounts for a great number of advances in building technology. Religion's central role in most civilizations usually ensured vast expenditures of resources, manpower, and talent on their structures.

Many of the great monuments of ancient societies still elude understanding. How were 50-ton sarsens transported to Stonehenge? And why? What did the bluestones signify? Across the Atlantic, Spanish conquistadores concluded that the 200-ton blocks at one Inca site had been positioned by magic. Huge monuments built without wheels or pulleys might suggest supernatural intervention to some, but the builders labored mightily to link heaven and earth. Egypt's greatest pyramids symbolized not only pharaonic power but also celestial stairways; Sumerian ziggurats, visible 25 miles away, were "hills of heaven."

In addition to overcoming logistical obstacles, builders of such monuments solved basic construction problems. Recesses and buttresses alternately lightened and reinforced the mud-brick ziggurats, for example. Far more advanced, the pyramids at Giza employed corbeling to deflect the downward thrust of thousands of tons of limestone from the internal passages and burial chambers.

As freestanding temples evolved, they usually took the simplest of structural forms: beams and columns. The gleaming marble temples of Greece's Golden Age derived from simpler timber structures. But on a large scale, unreinforced masonry

ST. PETER'S BASILICA, VATICAN CITY

204

PRECEDING PAGES: Sandstone uprights still shoulder a brief run of lintels along Stonehenge's northeastern flank. Nearly seven tons each, the lintels inspired the site's Saxon name, "place of hanging stones," and once came within an inch of forming a perfect, level circle. Roughly contemporaneous with the Giza pyramids, Stonehenge appears to share much more with classical Greek temples, especially post-and-lintel construction. This was an adaptation of a building technique used for timber.

At Stonehenge, mortise-and-tenon joints secure lintels to supporting sarsens, while tongues similar to toggle joints link each lintel in the outer circle to its neighbor.

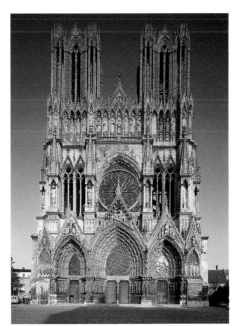

REIMS CATHEDRAL, FRANCE

THE TEMPLE OF AMUN, KARNAK, EGYPT

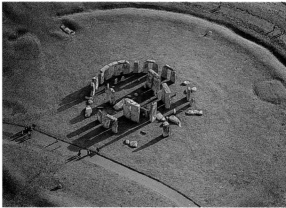

STONEHENGE, ENGLAND

couldn't bear up to the tensile stresses of beam-and-post construction.

Masonry could be extended over huge spaces, though, if curved. Roman builders created the masonry dome, among the most versatile of spanning structures. They took advantage of local materials such as volcanic rock and sand to make their supple concrete. The pride of an empire at its peak was reflected in the great dome of the Pantheon, the largest in the world until modern times.

Elsewhere, local resources also shaped structures. Readily available hardwoods gave rise to Japan's massive Buddhist temples; Russia's easily curved timbers made possible its bulbous onion domes.

In the West, church building took a cosmopolitan turn after Crusaders introduced the Middle Eastern pointed arch to Europe. Elongated into the Gothic vault, it rapidly replaced the semicircular arches of Romanesque churches, and radically reduced outward thrusts.

Gothic cathedrals reflected a civilization in transition as the Dark Ages receded: On the wings of buttresses, their soaring vaults literally reached toward the light. But every civilization has had a similar need for self-expression. Time after time, their religious yearnings have inspired mortals to overcome building challenges and respond to the spirit of their age.

205

Pyramids

More than the sum of its 2.3 million blocks, the Great Pyramid at Giza still guards many secrets of its construction.

E gypt's largest pyramids are the most massive monuments ever built. The base of the Great Pyramid alone could contain the cathedrals of Milan and Florence and London's St. Paul's, along with St. Peter's Basilica and Westminster Abbey.

More than by their scale, however, the pyramids awe by their implausibility. Most were built between 2700 and 2100 B.C. by a civilization that labored without benefit of the wheel, the pulley, or a metal tougher than copper.

Tombs for the Old Kingdom's pharaohs, the pyramids may also have served as temples to the sun god Ra. In practical terms, they were also vast public works projects that employed hundreds of thousands of conscripted farmers idled by the seasonal flooding of the Nile. Brute manpower emerges as the central fact of the pyramids' success.

NAPOLEON AND MEMBERS OF HIS EXPEDITION AT GIZA, 1799

Pyramids

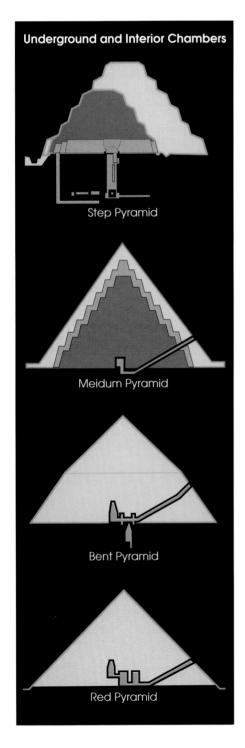

Underground and Interior Chambers

Step Pyramid

Meidum Pyramid

Bent Pyramid

Red Pyramid

Some 80 pyramids still stand, most where desert encroaches upon the Nile's fertile west bank. Even more ancient, benchlike mastabas and other brick-covered royal tombs of the Early Dynastic period (circa 2920–2575 B.C.) signal the pyramid's origin. From simple earthen mounds covering crude burial pits to squared-off structures of sunbaked brick concealing elaborate underground compartments, these tombs evolved into likenesses of grand dwellings.

But nothing that came before it compared with the tomb built about 2630 B.C. in Saqqâra for the pharaoh Djoser by his chancellor and architect, Imhotep. Out of the "material for eternity," as the Egyptians called it, he planned the world's first monument made entirely of stone. The design was, initially, less venturesome: Over a subterranean warren of chambers and passageways, Imhotep set a solid, square mastaba, its sides measuring some 200 feet, its tapering walls rising to 26 feet.

Then, for reasons unclear, Imhotep began tinkering with his creation, enlarging it and building it higher until it was some 200 feet tall, with a 400-by-350-foot base. Courses of rock laid atop the structure radically changed it. No longer a mastaba, the limestone-cased Step Pyramid introduced a new structural shape.

Djoser's monument might have remained Imhotep's folly had not the steps given physical form to religious thought. Entitled in the afterlife to his place among the "imperishable" circumpolar stars, the pharaoh required a means of ascending to the heavens. The steps of his tomb became his celestial staircase.

Like the Step Pyramid, many later pyramids were crisply aligned with the four points of the compass, perhaps in relation-

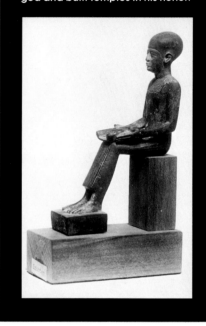

Deified after death, royal architect Imhotep originated monumental stone construction, built the first pyramid, and may have engineered ancient Egypt's irrigation. The Greeks thought him a healing god and built temples in his honor.

ship to the never-setting polar stars. But stepped pyramids were built only until religious beliefs began to shift, around the beginning of the Fourth Dynasty in 2575 B.C.

About 33 miles south of Saqqâra, the last of the great stepped pyramids, at Meidum, incorporated the essential transition. Planned as a small stepped pyramid and completed as a larger, eight-step structure with sides inclining at a 75-degree angle, the Meidum pyramid was later converted into a true pyramid. Its steps were filled in with local stone and its faces mantled with dressed limestone. More truncated tower than pyramid today, the Meidum structure lies surrounded by 250,000 tons of limestone blocks—the rubble of its outer layer, left by later builders who broke up

Harnessing simple technology for large-scale building, Imhotep gave the Step Pyramid (left) its final form after five major alterations. From a rectangular base, more than a million tons of limestone rise to a height of 204 feet.

Uncloaked, the Meidum pyramid (below, top) reverts to a stepped shape without its limestone sheath. The Bent Pyramid (middle) retains more facing than any other pyramid. The first true pyramid, the 350-foot-high Red, or North, Pyramid (bottom) derives its name from the ocher hue of local stone.

its casing for use in other structures.

The Meidum pyramid's final angle of slope was 52 degrees, close to the typical batter of most later pyramids. The Bent Pyramid was begun at a slope of 54 degrees, but about halfway up this was changed to 43½ degrees. The site chosen was too weak to support the weight of the massive structure. When settling and cracking began during construction, the Bent Pyramid's builders retreated to the more conservative angle, which reduced the volume of stone. And the next pyramid, the North, or Red, Pyramid at Dashûr, was the only pyramid built entirely at the gentler, safer angle of 43½ degrees. It was also the first true pyramid to be completed.

Other important changes occurred as mastaba became pyramid. The Step Pyramid was built of small stone blocks resembling the mud bricks used in mastabas. But for the pyramids at Meidum and Dashûr, huge blocks were quarried and dragged on sledges to building sites.

With time and trial, the builders also became adept at corbeling—building false arches and vaults by overhanging each successive layer of stone inward over the one below it until two sides meet. Corbeling allowed the Egyptians the option of building apartments within the pyramid, instead of having to bury them underground.

The Egyptians mastered enormous problems of monumental construction one by one, setting the stage for the grandest pyramids of all, at Giza.

The Great Pyramid

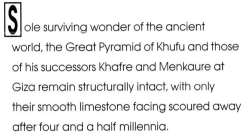

S|ole surviving wonder of the ancient world, the Great Pyramid of Khufu and those of his successors Khafre and Menkaure at Giza remain structurally intact, with only their smooth limestone facing scoured away after four and a half millennia.

Building for the ages was not sufficient for the Egyptians, though. Preoccupied with their pharaohs' afterlife, they aimed for eternity. Scholars theorize that by the beginning of the reign of Khufu, about 2551 B.C., the ascendancy of the sun cult of Heliopolis as the official religion had heightened the pyramid's symbolic importance: It reproduced in stone the sunbursts that pierced the clouds and projected pyramids of light across the Nile.

The rapid evolution of the pyramid under Khufu's father, Snefru, gave the Egyptians the practical experience to build large pyramids penetrated by intricate internal structures, and to undertake construction at what became the ideal 51°52′ slope. Still, Khufu's monument involved mind-boggling leaps of scale: From a base covering 13.1 acres, the Great Pyramid soared 481 feet (about 30 feet have since been lost), its 2.3 million stone blocks averaging 2.5 tons each.

Sited on a rise, Khafre's pyramid (center) only appears larger than Khufu's (background).

A winding shaft (1) linking Khufu's Grand Gallery (2) with a descending passage (3) allowed those who sealed the upward passage (4) to exit after the burial.

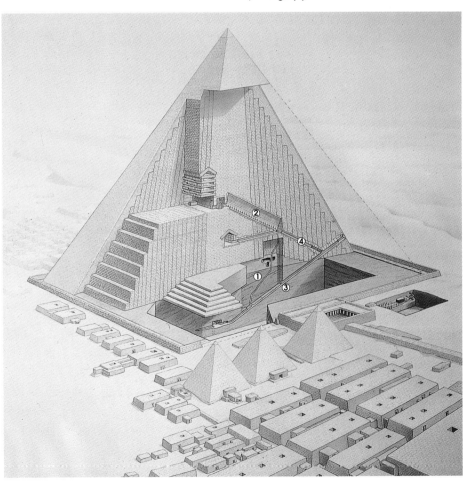

The Great Pyramid

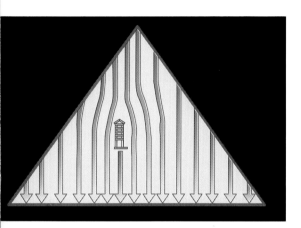

To deflect the load of 400 feet of stone above it, the burial chamber was buffered with five weight-relieving chambers. Four have flat roofs; the topmost has a pointed saddle roof that redirects downward thrust into side walls.

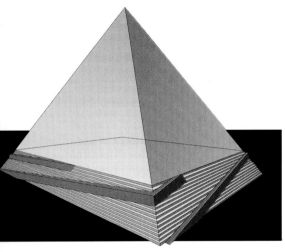

Ramps spiral upward from each corner, according to one of many theories of pyramid construction. Proponents note the ease of extending just the ramps' upper ends. But critics point out that this system would have concealed the structure's corners and edges, which surveyors needed for their calculations.

With help, a visitor (upper right) scales the pyramid's huge blocks in 1911.

Two resources were in abundant supply: time and manpower. According to Greek historian Herodotus, preparing the limestone plateau at Giza took a decade; building the pyramid, two more. After the site was chosen, workers stripped it of sand and leveled it, probably by digging, inundating, and gauging water levels in trenches. Whatever method they used bested most present-day construction planning: The pyramid's four corners vary in level by no more than an inch.

Surveyors using string and perhaps stars for sighting laid out four straight sides 750 feet long. These differ by less than eight inches. Equal precision prevailed as the royal astronomer and his assistants oriented the sides to the four cardinal points.

An 800-yard causeway of carved and polished stone built from the harbor was "no less a feat . . . than the pyramid itself," wrote Herodotus. While casing blocks of creamy Tura limestone and 50-ton granite slabs from Aswan were floated in on barges, local quarries primarily yielded the interior blocks of limestone.

To quarry the stone, workers made ingenious use of crude tools. Wedge sockets were picked out with copper tools. Wooden wedges, inserted in the cuts and pounded with wooden mallets, were wetted until they swelled and cracked the limestone. Probably, fires burned granite to crack off its weathered face; dolerite balls pounded it away from the rock mass; then quartzite, sand, and water polished it.

Tura limestone was entrusted only to the most skilled masons. With copper saws and chisels they planed the facing blocks to silkiness, and the joints so precisely that a postcard will not fit between them.

About 4,000 skilled masons were employed full-time at the quarries and the building site. As many as 30,000 unskilled workers might have labored seasonally. They built temporary ramps of rubble and clay up the pyramid's sides, slicked the ramps' mud surfaces with water for easier hauling, positioned timbers to provide a firm bedding, and hauled the stones into place. Wooden rockers and levers helped to position the stones, which slid into place over a fine layer of liquid mortar. The pyramid rose course upon steplike course, each from the inner core outward, perhaps as a series of stepped and angled buttress walls. Finally, the facing stones were set, at the 51°52' angle, starting at the capstone.

Externally, the Great Pyramid was similar to later structures; internally, its arrangement of passages and chambers was unique. Near the base, a traditional north-side descending corridor burrowed to an underground chamber. An ascending corridor forked from the descending passage to a level passage into another apartment, known as the "Queen's Chamber" although no queen was buried there.

The ascending passage was extended into the 153-foot-long Grand Gallery, a rising corridor with a 28-foot-high corbeled roof. Beyond lay Khufu's final resting place: a simple room roofed by 400 tons of granite. Above that, relieving chambers were built to deflect the enormous load bearing down from the pyramid's vertical axis.

Khufu's sarcophagus, an inch too wide for the entryway, preceded the pharaoh into his burial chamber during construction. So did several enormous granite portcullises that would plug the pyramid's entrance after his death in a futile attempt to foil tomb robbers and souvenir hunters down through the millennia.

An aerial view of the Great Pyramid shows it rising 450 feet at an angle of 51°52'. Originally some 30 feet higher, it was not exceeded in height by another structure—England's Lincoln Cathedral—until the early 14th century.

Temples

Massive columns rise in the Hypostyle Hall of the great temple complex at Karnak, Egypt, erected about 1300 B.C.

Since ancient times, societies have sought to understand and to live in harmony with the universe as they perceived it. Many of their structures reflect this age-old quest.

The great ziggurat at Ur, in present-day Iraq, suggests the cosmic mountain central to ancient Middle Eastern mythology. Brick facings eight feet thick sheathe the mud-brick core. Three tiers draw the eye upward to the terrace, once the site of a temple sacred to the moon god, Nanna.

The siting of the great stone circle at Stonehenge, in England, testifies to the importance of celestial events—such as the summer solstice—in the lives of early agricultural people.

In Nara, Japan, eighth-century builders tested the limits of timber-frame construction to house the world's largest bronze statue of Buddha in the great hall of the Todaiji temple complex. Thus they conveyed the enormous spiritual impact of Buddhism on their culture.

Temple builders responded to the need for spiritual expression not only with structures on a massive scale, but also with works of beauty that were in harmony with the natural surroundings. They pushed their skills to the utmost to create structures that were meant to last. Dedicated to the worship of gods, such buildings were worthy of the finest work and the best technology that builders everywhere could offer.

RECONSTRUCTED ZIGGURAT AT UR, FIRST BUILT AROUND 2100 B.C.

The Parthenon

The best-known Greek temples were built in the period between the defeat of the Persians in 480 B.C. and the death of Alexander the Great in 323 B.C. Perhaps the most famous is the Parthenon, which incorporates refinements in structure and design commonly used in fifth-century Athens.

The elegance of Greek temples was achieved through subtle, lucid design. Stability depended on sturdy underpinning. Builders normally carried a foundation of carefully coursed masonry—usually of limestone—down to bedrock, often to a depth of many feet. On this rested a platform made up of three or four courses of

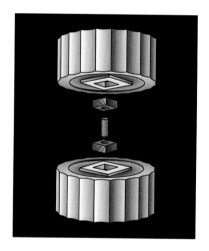

marble, held together in both the horizontal and vertical planes by iron dowels embedded in lead. Such construction was resistant to the earthquakes that occasionally shake this seismically active region.

The masonry of the Parthenon, like that of other ancient Greek buildings, was laid without mortar. As shown in the diagram above, individual columns were usually constructed of a series of drumlike units centered by socketed dowels. After the drums were carefully joined together, the surface of the column was channeled by master carvers.

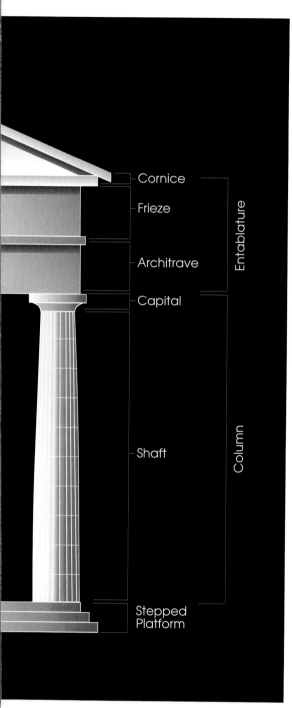

Cornice
Frieze
Architrave }
Entablature
Capital
Shaft
Column
Stepped Platform

Columns of Doric temples such as the Parthenon stand on a platform of stone slabs resting on a deep stone foundation. The capital transmits the load of stone above it to the shaft for conveyance to the platform and into the earth.

A masterpiece of engineering as well as architecture, the Parthenon has withstood wars and earthquakes. The elegant temple is basically a simple structure of vertical supports and horizontal top and bottom stones.

Pentelic marble tools to a very smooth surface and allows sculptors to execute fine joints. Deft chiseling of adjoining blocks eliminates uneven patches. Metal cramps link the blocks together, as shown in this simple illustration.

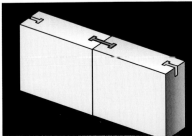

Chinese Timber-framing

A tiny timber hall in the Wutai Mountains southwest of Beijing, the Nan Chan temple dates from the eighth century, making it China's oldest existing wooden structure.

China, like Greece, endures frequent earthquakes, and centuries of experience led its builders to develop design and engineering practices that took seismic activity into account. To counter horizontal movements, builders devised flexible structures that moved with the earth, not against it. They relied on the integrity of their framework. Non-load-bearing walls in the structure might fall, but the frames would remain.

The principal building material was timber. Many kinds of wood were used, and those with excellent tensile strength, such as cedar, were highly prized.

Architecture entered a glorious age during the Song Dynasty (A.D. 960–1279). Builders in this period used a system of construction that had been developed several centuries earlier and that was based on time-tested rules. These rules, along with definitions of terms and other specialized data, were published in manual form in A.D. 1103. Song officials used the *Yingzao fashi* (Building Standards) to determine design and structural requirements and to estimate material and labor needs.

In typical structures, a raised platform rested on a tamped-earth foundation. The platform formed the base for a timber post-and-lintel frame, which supported a series of brackets that in turn supported a pitched roof with overhanging eaves.

The overall layout of a building, regardless of its size, was determined by the number and placement of standardized modular units called *jian,* or bays. Such a modular system facilitates prefabrication.

The roof was the most striking feature of a Chinese temple or hall. For protection against the elements, the roof extended well beyond the building's foundation. The traditional Chinese builder could produce

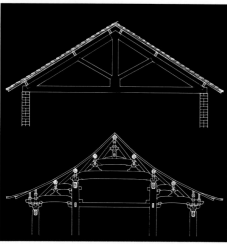

In conventional Western architecture, a triangular truss dictates the straight pitch of the roof (upper diagram). In China, equally simple components are used. As a timber skeleton of columns and beams rises toward the ridge (lower), the lengths and widths of its parts are adjusted so that a series of unevenly graduated steps underlies the purlins and rafters supporting the roof. Bracket sets on the columns help accentuate the curvature of the roof at the overhanging eaves.

Japan's Hall of the Great Buddha (opposite), in Nara, is perhaps the world's largest timber-frame building; some of its pillars rise more than 100 feet above the temple platform.

Bracket sets distribute the weight horizontally and redirect the outward thrust of the roof down through the columns.

a roof of any size and curvature simply by manipulating the lengths and widths of posts and crossbeams in the timber skeleton rising toward the ridge.

Built in 1056, the Yingxian Pagoda in the northern province of Shanxi is China's oldest surviving pagoda built entirely of wood. Its multiple levels of wooden bracing interlock to give stability. Intricate

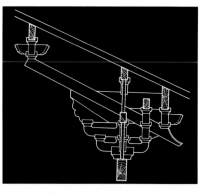

A slanted bracket arm, or *ang,* works like a seesaw. Roof load bears down on its upper end; its lower end rises to support the eave.

Chinese Timber-framing

patterns of brackets and beams create interior space sufficient to house religious images. Rising 183 feet, the five-story tower is a catalog of structural forms; 56 variations have been counted in the brackets alone.

The ingenious Chinese system greatly influenced building practices in a vast area of Asia. Japan's Hall of the Great Buddha in Nara, for example, was inspired by Chinese structures. Timber-framing is still carried on in parts of Asia today.

Buddha images gaze in the four directions on an upper level of China's Yingxian Pagoda. Earthquakes, floods, and human violence have left their marks on the 11th-century temple (above), but its sturdy frame stands firm.

Observatories

O n Salisbury Plain in southern England rise the ruins known as Stonehenge. The name, dating from medieval times and meaning "place of hanging stones," refers to structures that give the site its unique character: stone uprights with lintels.

More than 600 great stone circles dot the British Isles, but none was as carefully or as elaborately constructed as Stonehenge. Archaeologists theorize that it was a kind of temple where solstice rituals were performed. The structure's solar orientation would have made it easy for priests and shamans here to pinpoint the solstices and to plan ceremonies connected with the passage of the seasons.

Through centuries of speculation, Stonehenge was attributed to the Romans, the Druids, even the mythical Merlin. But in fact, it was built by prehistoric farmers.

Using deer-antler picks, the first builders dug a circular ditch in the chalky soil around 3000 B.C. They made a northeast-facing entranceway, flanked by two stones. Another pair of stones was set outside the entrance; of these, only the so-called heel stone remains. Centuries later, the entrance was shifted slightly so that

To build the familiar linteled-sarsen structure, stoneworkers first shaped the massive sandstone slabs by battering them with simple stone tools, leaving knobs, or tenons, at the top of each upright to fit hollowed-out mortises in the lintels. They curved the lintels slightly so that when linked together with tongue-and-groove joints they formed a continuous circle.

Workers hauled the stones to ramped pits and hoisted the uprights into position. Then they slowly lifted the lintels by placing layer upon layer of timber under them. Finally, they edged them sideways onto the uprights.

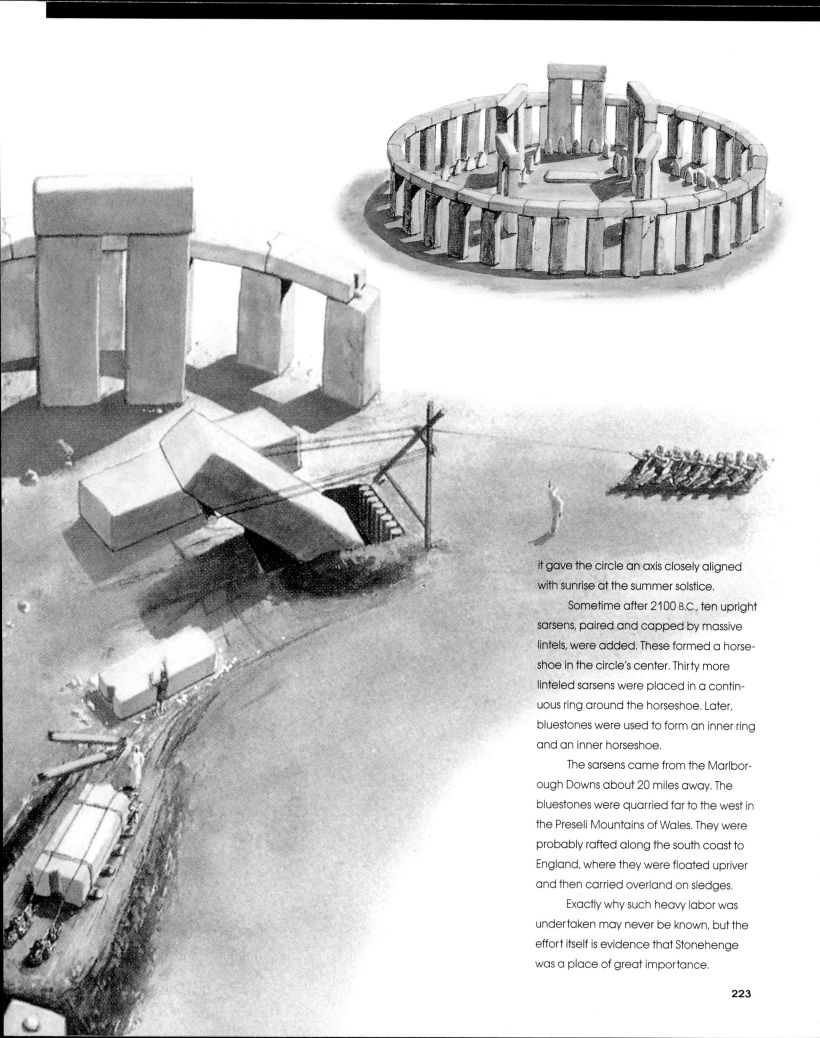

it gave the circle an axis closely aligned with sunrise at the summer solstice.

Sometime after 2100 B.C., ten upright sarsens, paired and capped by massive lintels, were added. These formed a horseshoe in the circle's center. Thirty more linteled sarsens were placed in a continuous ring around the horseshoe. Later, bluestones were used to form an inner ring and an inner horseshoe.

The sarsens came from the Marlborough Downs about 20 miles away. The bluestones were quarried far to the west in the Preseli Mountains of Wales. They were probably rafted along the south coast to England, where they were floated upriver and then carried overland on sledges.

Exactly why such heavy labor was undertaken may never be known, but the effort itself is evidence that Stonehenge was a place of great importance.

223

Observatories

At the 18th-century observatory in Jaipur, India, the shadow cast by the gnomon (below) of a gigantic sundial on quadrants of a circle permits observers to track the sun's movement and thus determine the time of day. Nearby, a concave hemisphere called a *jai prakas* (right) locates the sun's position using the shadows made by two intersecting wires.

For millennia, the authority to rule was linked with the ability to foretell the future. Power accrued to those who could predict eclipses, phases of the moon, the reappearance of certain stars, and the winter and summer solstices, and who knew the proper days for performing rituals, for waging war, for planting and harvesting.

To have such knowledge required making regular celestial observations and carefully recording them. Before the development of the telescope, sighting techniques relied on the naked eye. Often, observers were aided by structures or instruments of considerable size. These might have a solar orientation or celestial alignments with doorways, windows, or markers.

A medieval stargazer, Prince Ulugh Beg of Turkestan, built an observatory in Samarkand that was considered a wonder of the world. One of his measuring devices was said to rival Hagia Sophia in height.

In 18th-century India, Prince Jai Singh II constructed an observatory at Jaipur that held about a dozen stone instruments, including an enormous sundial. The sundial's 90-foot gnomon has a hypotenuse paralleling the earth's axis. On each side is a quadrant of a circle, paralleling the plane of the Equator. At sunrise the gnomon's shadow falls on the highest point of the western quadrant; it descends until noon, then ascends the eastern quadrant.

Such simple masonry instruments provided sufficient data until relatively recent times, when they gave way to the exquisite precision of modern telescopes and atomic clocks.

Unlike the modern observatories it resembles, this temple (above) at Chichen Itza in Yucatan housed no scientific instruments; it honored a Maya god. Windowlike openings align with the setting sun at the equinoxes, and with Venus, the planet associated with war and death.

Angled like the gnomon of a sundial, the 500-foot-long shaft of the McMath Solar Telescope at Kitt Peak National Observatory, in Arizona, parallels the earth's axis of rotation. Large mirrors at the top constantly track the sun from east to west, reflecting its light down the shaft to an underground observation room.

Its square window aligned with the polestar, this 30-foot tower (right) at Kyongju, South Korea, has stood since the seventh century. A wooden platform at the top supported instruments for tracking celestial objects.

225

Domes

Eye on the Roman sky, the Pantheon's 27-foot oculus beams sunlight on visitors 143 feet below.

Forms that suggest heavenly spheres, domes answer the earthly challenge of enclosing large spaces without internal supports. For almost two thousand years, domes have held the record for spanning the greatest interior spaces.

For much longer, they have curved over ancient mounds, tombs, and small temples. Igloos, hogans, and yurts still depend on the form. Conceptually, it is created by rotating an arch around a vertical axis and providing continuous support along the resulting perimeter. This continuous support accounts for the efficiency of domes, which send their loads downward along curving paths to their foundations.

In domes, structural forces exert powerful outward thrusts that prevented early builders from doming any but the most basic of freestanding structures. Undaunted—and drawn to arching shapes in their bridges, aqueducts, and other structures—Roman builders freed the dome from its humble past. To do so, they used concrete, a strong new material composed of high-grade cement with aggregates of stone and brick fragments. By the second century A.D., concrete had superseded brick and stone for spanning large buildings.

So momentous was the transition that historians refer to a "Roman architectural revolution" wrought in concrete. It occurred at the empire's zenith, a time of building on a monumental scale.

Roman genius transformed interior space, and the span of one of its greatest achievements, the Pantheon, was unsurpassed until the 19th century. Still, the Romans left a few problems for later builders. Byzantine architects solved one by perfecting a method for placing domes atop square or rectangular structures.

Other innovations rode a surge of renewed interest that began with the Renaissance. In his design for the cupola of the Florence Duomo, Filippo Brunelleschi divided the dome into inner and outer shells.

Lighter and stronger materials—cast iron, reinforced concrete, steel, inflatable man-made fabrics—and advances in physics and mathematics have simplified dome building and shifted it toward less sacred purposes. Today's huge domes, such as the Louisiana Superdome, are monuments to secular society.

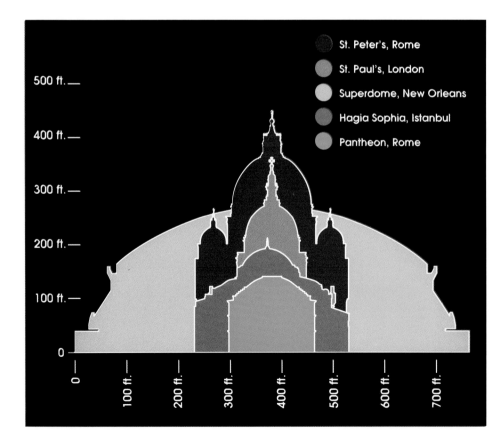

St. Peter's, Rome

St. Paul's, London

Superdome, New Orleans

Hagia Sophia, Istanbul

Pantheon, Rome

The Pantheon

Dedicated in its day to "all the gods," the Pantheon today honors an unknown Roman architect whose vision created one of the most influential buildings of any era. Structural soundness and its consecration as a church in 609 have combined to keep the Pantheon in use for nearly 1,900 years. Though no longer a church, it remains the most outstanding of all ancient structures still largely intact.

Roman engineers had built domed buildings before. But the dome of the emperor Hadrian's new Pantheon, begun in 118 and completed a decade later, was to be far larger than any other. It demanded a formidable support system. Possibly building on the site of an earlier temple, workers began with a 24-foot-wide circular foundation wall.

Above it, they built 20-foot-thick support walls: effectively a cylinder to contain the dome's outward thrust. Though thick, the encapsulating walls are riddled with voids; inside them, a complex web of vaults and arches channels the vertical load from the dome down to eight massive piers.

The whole brick-faced concrete structure uses lighter aggregates as it rises, lightening the load on the supports below. At the crown of the dome, the monolithic shell tapers to 59 inches thick.

Though chunky compared to the modern, machine-mixed variety, Roman concrete was so supple that, when wet, it could be curved and shaped at will over formwork. Cemented by excellent mortars of volcanic sand, it contained varying mixes of aggregates: stone, volcanic rock, brick, even rubble from demolished buildings. These turned concrete airier or weightier, allowing builders to modulate the density of their structures. Rising layer by horizontal layer, the Pantheon's concrete lightens, reducing loadings in the material to roughly half that of heavier aggregates. Mixes favoring heavy basalt yield to lighter brick shards, then yellow tufa, then airy pumice.

And then air itself. The dome culminates in a 27-foot open oculus, or eye, lined with brick and sheathed in gilded bronze, that floods the Pantheon with light and suggests the truest legacy of its architect: making the inside, not the outside of the building the focus of attention.

Most earlier monumental architecture had been concerned with outer, sculptural form. Interior spaces (except in Roman buildings) were often broken up by masses of columns. The Pantheon's architect achieved just the opposite effect. In a street view, the dome is undistinguished, and the building gives few hints of the soaring spaces and grand proportions that still awe visitors stepping from the gabled porch into the rotunda.

Because the distance from floor to oculus, 143 feet, is the same as the interior's

diameter, the hemispheric dome would touch the floor if it were extended to a complete sphere. Though the dome's dimensions made that spatial harmony possible, other factors enhance the effect. The large hollows between each of the eight supporting piers extend the eye horizontally, while niches and coffers divide lower cylinder and upper hemisphere into a series of vertical zones, giving full play to a new realm of spatial nuance.

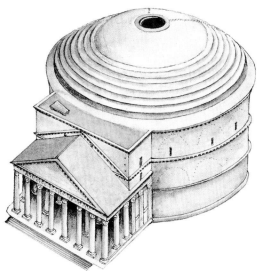

Overlapping layers of concrete (left) create concentric stepped rings that thicken the dome's base, as in Roman arch construction. The rings also serve to lessen stress. Lead roofing replaced the original bronze in the 1600s.

The dome may have taken shape over timber formwork, as in the cutaway below. On the dome's interior, five rings of square coffers in a wafflelike pattern, diminishing in size and depth, create illusions of space.

At far left, two levels of large brick barrel vaults channel the dome's load to piers below—and show the complex construction of the walls.

Hagia Sophia

As ethereal inside as it is massively material outside, Hagia Sophia (opposite) depends on a well-concealed support system.

Billowing semidomes at east and west provide partial support for the central dome. Buttresses north and south foreshadow Gothic architecture. Minarets appeared in Muslim Istanbul after the Ottoman conquest in 1453.

A golden dome suspended from heaven," wrote the historian Procopius of Hagia Sophia, the principal church of the Byzantine Empire, not long after its completion in 537. The greatest of Byzantium's domed structures, and for nine centuries the world's largest church, Hagia Sophia reflects, as the Pantheon does, an empire at its height and an emperor, Justinian, who pushed monumental building arts to new limits.

Hagia Sophia's chief designers were Anthemius of Tralles, a geometer, and Isidorus of Miletus, a natural scientist, known to their contemporaries as *mechanopoioi,* makers of machines. They carefully applied geometry and their observations of structural problems in earlier domes to their own project. During the five short years of

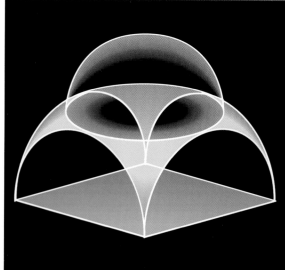

A hallmark of monumental Byzantine and later Ottoman architecture, pendentives may have been first used by earlier Roman architects for small buildings. Equilateral triangles shaped from a great sphere provide the form. Pendentives make the transition from a square to a circular plan by filling in spaces between the corners of the square and the dome's base. Hagia Sophia introduced their large-scale use.

Internal supports turn to aesthetic advantage as Hagia Sophia unfolds in bilateral symmetry. The great dome's support system consists of 40 ribs and piers that form four great arches joined by pendentives. To the east and west, semidomes of the same diameter as the central dome help to brace it, and the semidomes in turn open out to smaller half-domes. Colonnaded aisles and galleries are carefully integrated into the open, light-filled plan.

construction, Anthemius and Isidorus devised daring solutions to Hagia Sophia's unique building challenges. For instance, researchers have shown that the necklace of windows that helps create the illusion of a "suspended" dome is an expedient to prevent meridional cracks.

Perched atop four great piers, Hagia Sophia's brick dome appears to hang from the sky—a decided contrast to the Pantheon's, resting snugly in its massive supporting cylinder. The design, altogether new, fulfilled the quest to place a round dome on a square or rectangular base.

Early churches took two distinct forms: the rectangular basilica and the round, domed building. Justinian handed Anthemius and Isidorus the formidable task of wedding the two forms into a kind of domed basilica.

The building's overall floor plan encloses a rectangle some 220 by 300 feet. Under the central dome there is a 107-foot-square space defined by four immense piers of ashlar limestone, a local building material considered better able to handle the huge forces than brick-and-mortar construction. The piers, in turn,

support four giant arches, made of high-quality brickwork. The arches are joined over the piers by triangular spherical segments known as pendentives. This system forms the continuous support for the base of the dome.

Although its actual span is smaller than that of the Pantheon, Hagia Sophia appears larger because of the addition of half-domes and semidomes, set behind the arches at east and west. The builders may have intended these structures, as well as buttresses to the north and south, to absorb part of the dome's tremendous outward thrusts.

Substantial deformations of the main piers occurred even before the arches were completed. The frightened architects strengthened the massive mortared-brick buttresses outside Hagia Sophia's north and south walls. Even so, this didn't prevent the dome's collapse in 558, after two earthquakes. The dome that replaced it was 20 feet higher, soaring to some 180 feet. The second dome has also suffered partial collapses after earthquakes, but remains essentially in its original form.

Hagia Sophia

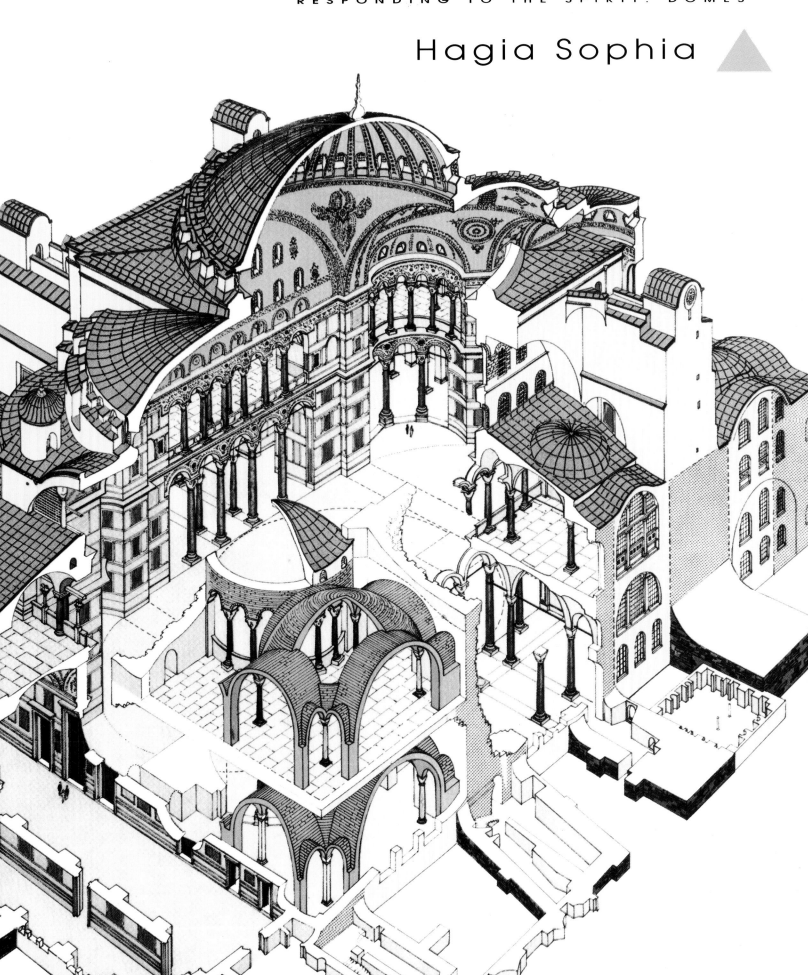

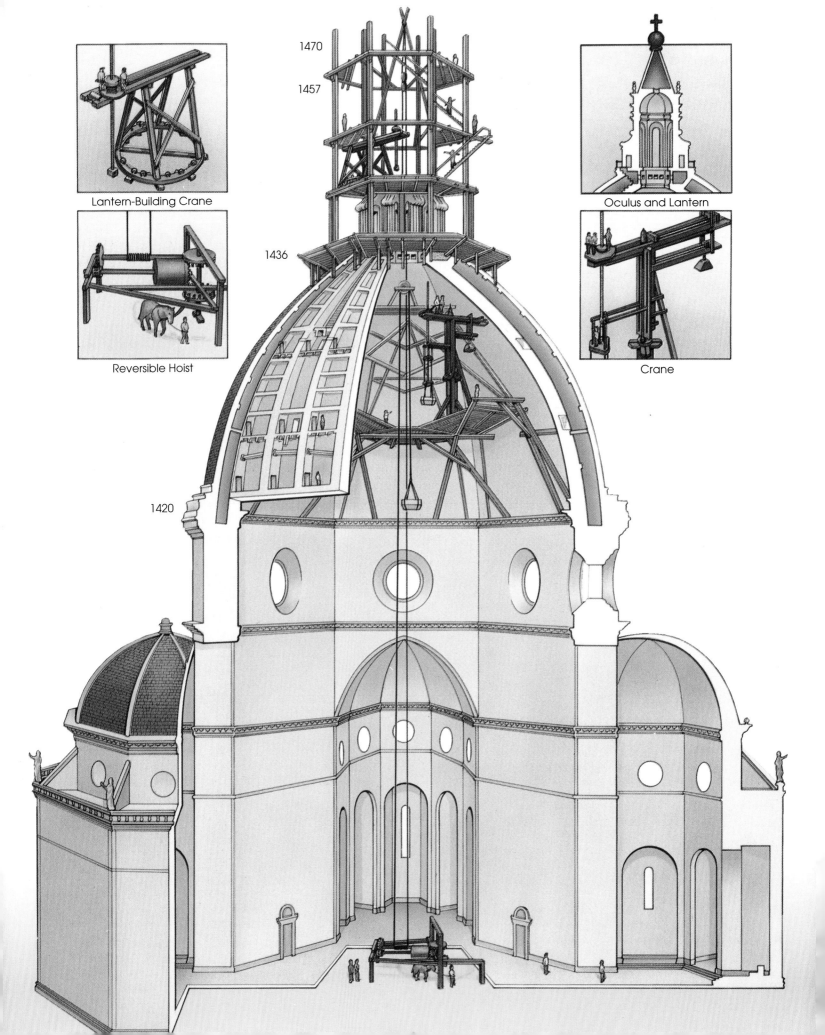

Lantern-Building Crane

Reversible Hoist

Oculus and Lantern

Crane

1470

1457

1436

1420

Santa Maria del Fiore Cathedral

Merging art and science into a standard for the future, Filippo Brunelleschi's dome for Santa Maria del Fiore, Florence's cathedral, boldly defined the Renaissance when it was finally completed in 1436.

Begun in 1296, the Gothic Duomo still lacked its cupola a century and a quarter later. The cruciform cathedral did, however, have a large altar area defined by eight walls and a drum taking shape above it; the original plan called for a soaring octagonal dome. But the plan was without precedent, and the problem of building a huge octagonal masonry dome seemingly without solution. The forest of timber trusswork required to erect a 25,000-ton octagonal dome could not be made secure. Moreover, wood was costly. The Duomo's guild council announced a competition for a viable design in 1418.

"Senza armadura": The dome could be built without wooden armature, Brunelleschi insisted. It would rise as a concentric series of self-supporting horizontal rings of brick and stone. When four masons were able to build a 12-foot-diameter model of his circle-within-an-octagon, Brunelleschi won the contract in 1420.

Brilliant as that solution was, other features of his design were revolutionary. One was the division of the cupola into a thick inner shell and a thin outer one. The outer dome, according to the architect, would protect the inner from wind and water. A staircase winds between them, easing inspection, repair, and, more recently, the curiosity of visitors.

Tapering as they rise, the domes are linked and reinforced by 24 vertical ribs of tan and gray sandstone. Eight ribs define the octagon's corners; the rest, smaller, are embedded in the sides. At the apex, 300

feet above the ground, the major piers converge around a *seraglio,* a stone enclosure encircling the 20-foot oculus. Above that, Brunelleschi's lantern was installed after the artist's death.

Another major innovation was the girdle of "chains," joined with iron cramps, that resist the dome's hoop tension. One, of wood, is partly exposed; six, of sandstone blocks, are completely hidden.

Brunelleschi's inventions made the job possible. His reversible hoist (opposite) raised heavy building blocks to a platform where a rotating crane put them in place. His lantern-building crane pivoted on rollers.

"Ample to cover with its shadow all the Tuscan people," the dome of Florence's cathedral spans 138 feet and rises to 300.

Saint Peter's Basilica

ppointed *capomaestro,* or chief architect, for Rome's new St. Peter's Basilica in 1546, the 71-year-old Michelangelo inherited the project from illustrious predecessors, among them Bramante and Raphael.

Michelangelo redesigned the dome with Brunelleschi's Florence masterpiece in mind. After his own death in 1564, only the drum was complete; when finished nearly three decades later, the dome, though altered, still reflected his mastery.

Directly over what many believe to be St. Peter's tomb, the double-vaulted brick dome rises to 452 feet at the top of its cross. The dome's diameter, 137 feet, is essentially the same as that of Florence. Four great piers, more than 60 feet wide, anchor its arches. Pendentives form a transition between the arches and the dome's circular drum. To counter hoop tension in the dome, caused by its tendency to expand, the builders wrapped two iron chains around it.

But neither chains nor the steep dome's lower supporting structures were sufficient to contain outward thrusts. After a century and a half, the ribbed dome had cracked badly, threatening to split open. In 1742, Giovanni Poleni, a professor of "experimental philosophy," was called in to consult. His pioneering application of structural mechanics to an architectural problem resulted in five more iron chains being wreathed around the dome. But Poleni also determined that the vertical cracks that segmented the dome into many arches posed no potential danger.

Michelangelo, nearly 90, presents a model of St. Peter's to Pope Paul IV in Domenico Cresti's 1619 rendition. No plans survived the artist. Words did: "It was not of my willing that I built St. Peter's," Michelangelo wrote, but rather "to the glory of God, in honor of St. Peter and for the salvation of my soul."

Rows of windows pierce 16 ribbed compartments as they curve toward the lantern. Though no longer visible from everywhere in Rome, as Michelangelo intended, the lead-sheathed dome still dominates the exterior of this tallest of all Renaissance churches.

Saint Paul's Cathedral

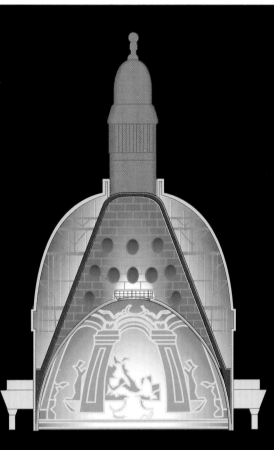

A result of London's Great Fire of 1666 was the rebuilt St. Paul's Cathedral. Completed in 1710, Sir Christopher Wren's dome made earlier dome construction techniques obsolete.

Wren admired St. Peter's and its double dome, but was alarmed by the structure's cracking. In addition, the rubble-filled piers of the new St. Paul's had started to spall even before the dome design was in hand. All these conditions demanded unprecedented lightness.

Wren achieved this by using three distinct structures: a lightweight inner, hemispherical dome of brick only 18 inches thick; a central brick cone also 18 inches thick, bearing at its top an 850-ton lantern; and an outer shell of lead-sheathed timber. A single iron chain resists spreading forces throughout the system.

"Reader, if you wish to see my monument, look around you." His epitaph refers to St. Paul's, but the legacy of Sir Christopher Wren (1632-1723) reaches to 100 buildings. Astronomer, inventor, and to Newton one of "the greatest Geometers of our times," he applied new structural concepts to architecture.

Catenary principles, developed by mathematician Robert Hooke, guided Wren to build his brick cone. The shape adopted by a chain hanging from two points, a catenary exerts entirely tensile, or pulling, force. But the same shape turned upside down is entirely compressive, the optimal state for a thin, load-bearing structure. An inverted catenary, Wren's cone is further stabilized and compressed by the lantern and reinforced by the chain at its base. These forces prevent the hoop tension that plagues St. Peter's and other great domes.

Spokes of light pour into the cathedral through 24 outer windows and 8 inner apertures.

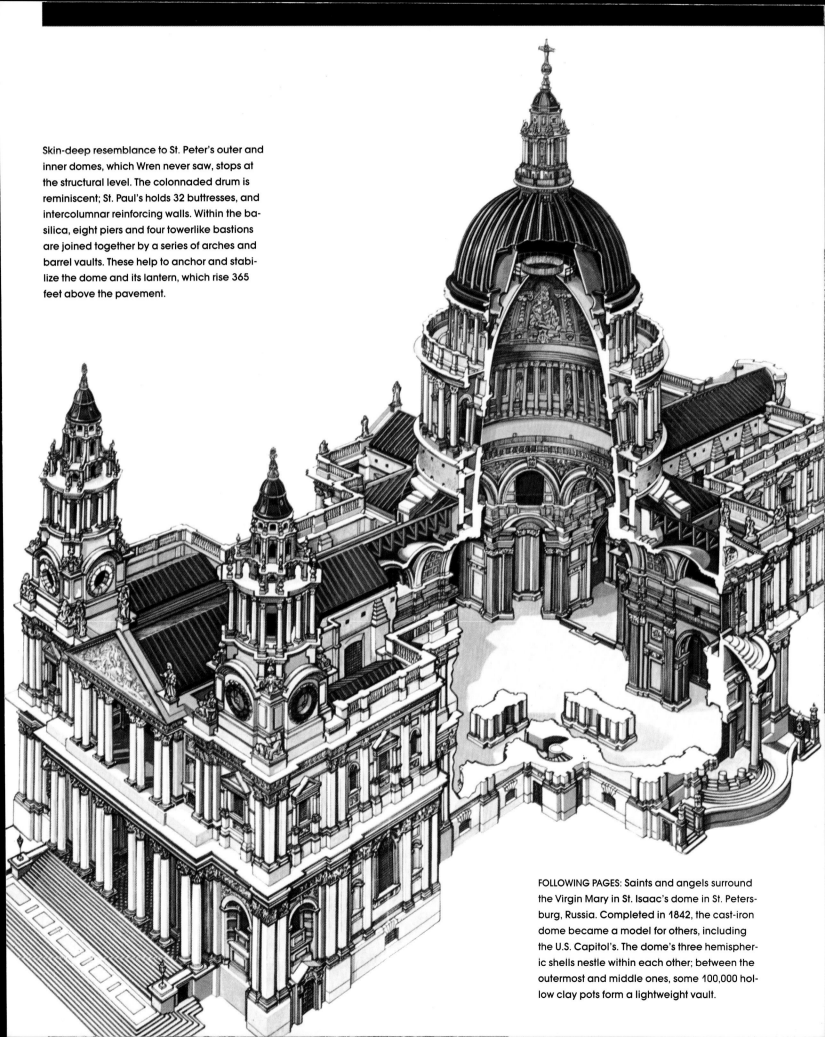

Skin-deep resemblance to St. Peter's outer and inner domes, which Wren never saw, stops at the structural level. The colonnaded drum is reminiscent; St. Paul's holds 32 buttresses, and intercolumnar reinforcing walls. Within the basilica, eight piers and four towerlike bastions are joined together by a series of arches and barrel vaults. These help to anchor and stabilize the dome and its lantern, which rise 365 feet above the pavement.

FOLLOWING PAGES: Saints and angels surround the Virgin Mary in St. Isaac's dome in St. Petersburg, Russia. Completed in 1842, the cast-iron dome became a model for others, including the U.S. Capitol's. The dome's three hemispheric shells nestle within each other; between the outermost and middle ones, some 100,000 hollow clay pots form a lightweight vault.

Gothic Cathedrals

Above lavishly sculptured arches, buttresses and pinnacles enhance the structural stability of Reims Cathedral.

The flowering of the high Middle Ages in France, from the mid-12th through the 13th centuries, was made possible by merchant wealth, religious fervor, and the organization of labor into specialized crafts.

In 1144, the completion of a new choir for the abbey church of St. Denis that incorporated ribbed vaulting and large stained-glass windows signaled the birth of the revolutionary new Gothic style. Abbot Suger, who commissioned the choir, believed in the spiritual power of light, and the quest for that light became the cathedral builders' greatest challenge.

Armed only with simple tools and geometric relationships, master masons formed high, pointed vaults and integrated them with a system of flying buttresses. Gothic cathedrals soon rose to unheard-of heights, and their radiant great rose windows became the emblems of the age.

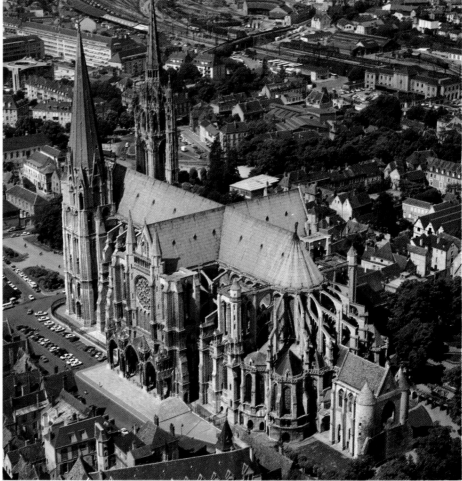

ABBEY CHURCH OF ST. DENIS (LEFT); CHARTRES CATHEDRAL

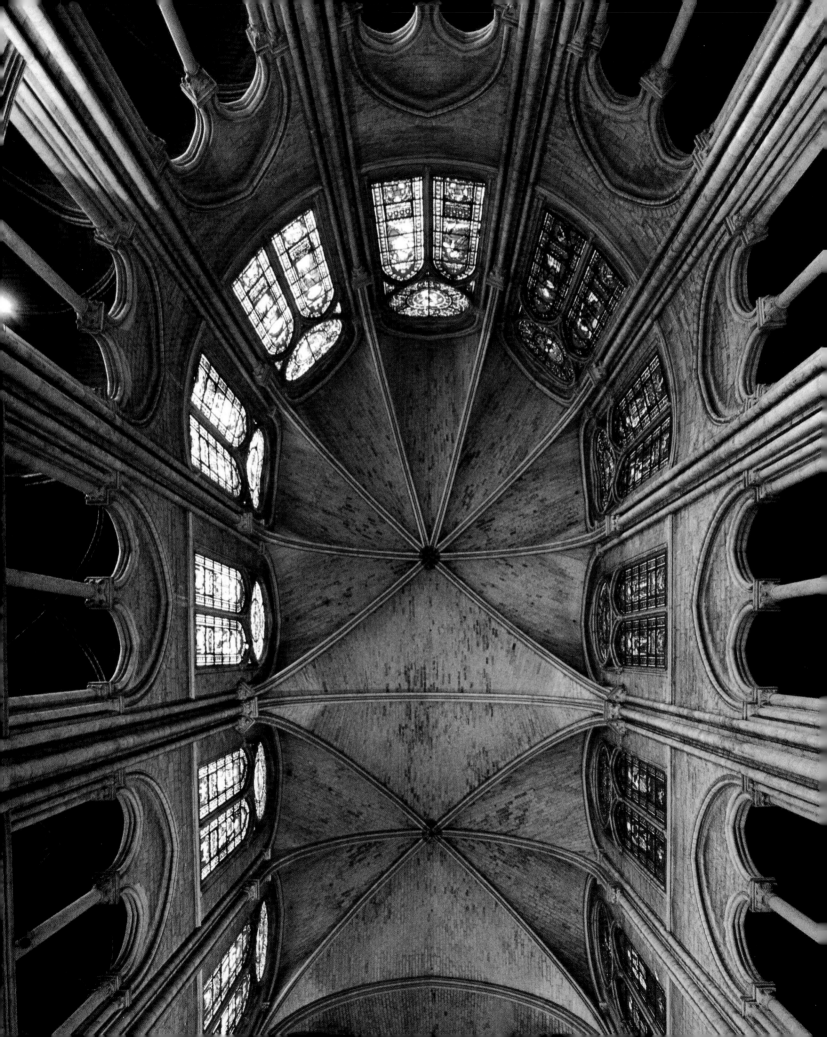

Notre Dame de Paris

In the spirit of competition that marked 12th-century cathedral building, the unknown Master of the Works at Notre Dame in Paris, begun around 1150, decided from the outset that it would be a quarter taller than its tallest predecessor. When the choir was almost complete, an even more daring decision was reached—to raise the nave even higher, to a height one-third greater than any other.

In this quantum leap, Notre Dame heralded the High Gothic style—and became the experimental workshop that set the course for its full flowering. The increase in height created problems that could not have been foreseen. In meeting the challenges,

the builders developed a structural vocabulary for all others to follow.

The original increase in elevation moved Notre Dame's clerestory windows so far from the floor that light barely reached it. When the walls were raised even higher to permit taller windows, another problem was introduced: higher wind speeds and much greater wind pressures than had been encountered before. A solution was found: The first flying buttresses were introduced around 1180. These revolutionary elements propped up the upper walls and braced them against the lateral thrust of the vaults and against

Six-part ribbed vaulting points to heaven in the choir of Notre Dame (opposite). Slender piers serve to carry its light weight, allowing the walls to turn into a panorama of glass.

Suspended in perfect equilibrium on a web of stone, the immense north rose window remains intact after 700 years, its intricately interlocking blocks so exact they ring when struck. Though individual blocks may be removed for repairs without collapsing the whole, only minor buckling has ever occurred.

Along with representations of saints and devils, medieval sculptors chiseled ordinary folk, like this ax-wielding workman (center), into the facade of Notre Dame.

Notre Dame de Paris

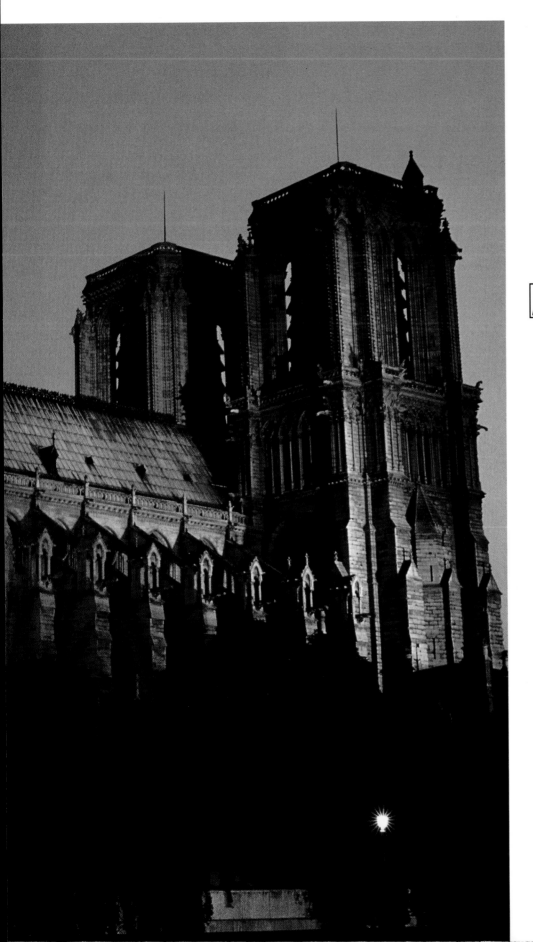

the much greater wind load. They were immediately adopted at other sites.

Modern stress analysis reveals some structural weaknesses in the original design of Notre Dame's buttresses. The builders reached this conclusion probably by observing cracked mortar—and in the 1220s they modified the structure, based on improvements developed in later building designs. Thin high fliers—members that soar up from the ground over the side aisles and gallery—were added.

At last, the stage was set for Notre Dame's glory—the creation of its great *rayonnant,* or radiating, geometric rose windows. Jean de Chelles, Master of the Works in 1250, decided to remove the plain walls of the transepts and replace them with walls of glass. Freed of bearing loads, walls were opened to a truly tremendous diameter. The north window with its rose, for example, occupies a rectangle 43 by 57 feet, supported only by a pair of flanking piers.

Stonecutters, whose skills had developed to an astonishing degree, cut hundreds of geometrically plotted interlocking blocks to form a framework for the glass. So perfect was the design of the north rose that its stone fillet has supported 1,300 square feet of glass for 700 years. The skill of Jean de Chelles, who plotted the geometry, and the work of the stonecutters were so precise that fewer than 20 of these windows were attempted in France in a hundred years—and none surpasses Notre Dame's in size.

Flying buttresses were added to Notre Dame around 1180, revised in the 1220s, and reconstructed in the 19th century. Built to counter wind loading on the nearly 100-foot walls, they inspired other builders to aim even higher.

247

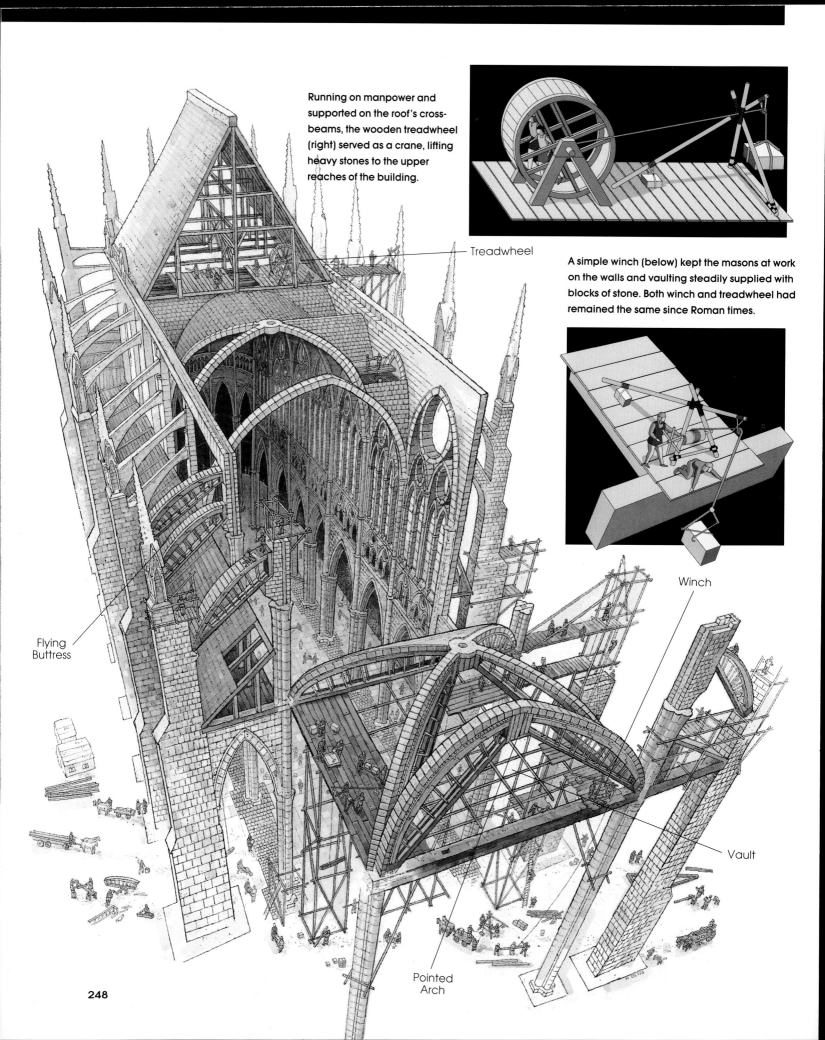

Running on manpower and supported on the roof's cross-beams, the wooden treadwheel (right) served as a crane, lifting heavy stones to the upper reaches of the building.

Treadwheel

A simple winch (below) kept the masons at work on the walls and vaulting steadily supplied with blocks of stone. Both winch and treadwheel had remained the same since Roman times.

Winch

Vault

Flying Buttress

Pointed Arch

Building Gothic Cathedrals

At the height of activity, the construction site of a Gothic cathedral swarmed with dozens of workmen, just like any high-rise site today. Members of craft guilds, the contracted workmen were organized into job teams and, unlike serfs, were paid for their labor.

Each site was supervised by a master mason, a master carpenter, and perhaps as many as 30 senior craftsmen. These specialists and some of their more skilled workers moved from job to job, applying lessons learned on one to the next.

Gothic engineering was not a science in the modern sense, but more an art. The master mason served as designer, artist, and craftsman, like those who worked under him. Armed with a level, a square, a triangle, a straight edge, measuring rods and strings, and a compass, he set the plan for the cathedral.

Building design must have been worked out either in drawings or in small-scale models. Once the floor plan had been established, the builders worked out full-scale details geometrically on a special plaster floor known as a "tracing floor." To minimize costs and to avoid massive re-construction if design problems occurred, the superstructure of a cathedral was often built one bay at a time.

Unlike the semicircular arch used in earlier Romanesque buildings whose height was fixed at half the span, the pointed Gothic arch allowed more flexibility. A range of heights was now possible for each span. Like the semicircular arch, the pointed arch was easily laid out using a segment of a circle.

Laid with a thin skin of stone over strong supporting ribs, much like an um-brella, Gothic vaulting was much lighter than Romanesque barrel vaulting, and focused supporting forces onto freestand-ing piers rather than massive walls.

Once the flying buttress was intro-duced at Notre Dame, intervening load-bearing walls were no longer required, and the walls could be opened with larger and more dazzling windows.

Construction began by laying out the foundation, which sometimes was as deep as 30 feet. On the foundation rested the plinth—the broad platform at the base of the walls, the footing of columns, and the solid base of the buttresses.

The master mason designed tem-plates for the masons cutting irregular or convoluted blocks, such as the bases of columns. The walls and buttresses were laid with ashlar, cut blocks of limestone. Once the plinth was laid, the walls (often made up of two skins of ashlar over a core of rub-ble, for economy's sake), the columns, and the buttresses were simply built upward.

Because of its expense, scaffolding was probably kept to a minimum. Medi-eval workmen consigned their souls to God and watched their step on withy plat-forms. Some scaffolding may have been pegged into the walls, rising as they did. A dangerous moment for the workmen came when the walls had reached their prescribed height, and the massive timbers for the wooden roof had to be manhan-dled into place, using blocks and tackle.

The roof was placed before the vaulting. Self-supporting, its crossties served as a platform for the lifting machinery that raised the vaulting stones into position.

Second only to the master mason in authority, the master carpenter designed and supervised the construction of the all-important temporary scaffolding, including

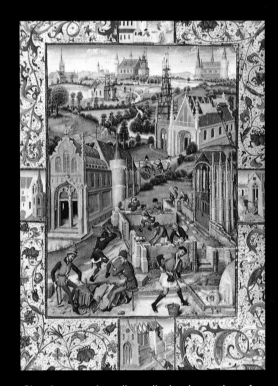

Churches sprout up all over the landscape in an Aus-trian illuminated manuscript, dated 1448. Colorfully illustrating the enthusiasm for building, it also high-lights the teamwork on cathedral job sites common throughout Europe. Well-dressed carvers (lower left) shape intricate blocks for either molding or window tracery. A common laborer, with holes in his stock-ings, hauls rubble for stone layers to put down on the walls between beds of ashlar. Another workman (lower right) mixes mortar.

Building Gothic Cathedrals

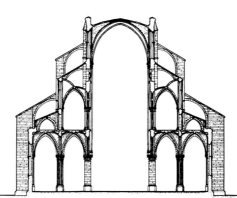

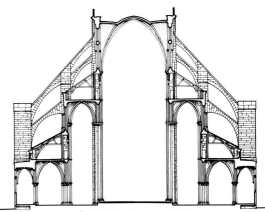

Laon, circa 1175

Notre Dame de Paris, circa 1180

Bourges, begun 1195

the heavy, braced centering frames that supported the arches and the ribs of the vault until the keystones were set and the arches complete. The master carpenter also knew geometry; his centering gave physical form to the master mason's geometrical planning, creating the exact angle and curve of every arch.

Once the centering frames had been positioned, masons laid the numbered, curved stones of the vaulting ribs, one by one. When they reached the point where the ribs intersected, the heavy keystone was hoisted and wedged into place.

The mortar took days or perhaps weeks to dry. The web of the vaulting, made up of much smaller and lighter stones, was then laid on top of centering laid over the ribs. Structurally, the web and the ribs each support themselves. The workmen probably stood on scaffolding, and the freshly mortared courses could have been kept from slipping by boards braced on edge between the centering frames. Similar methods are still used in Third World countries today.

Also busy at the cathedral site were craftsmen skilled in making and piecing

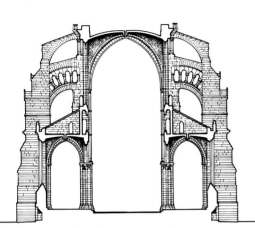

Chartres, begun 1194

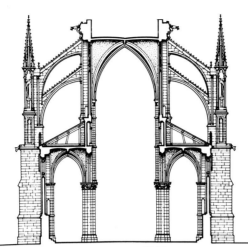

Reims, circa 1210

Amiens, begun 1220

As large towns vied for prestige, Gothic churches rose ever higher. From Laon, in 1175, to Amiens, begun in 1220, the height of the nave nearly doubled—to 138 feet (above).

At Amiens (opposite), soaring, repetitive vertical elements unite to evoke the sublime. While most cathedrals were constructed east to west, with the choir first, Amiens was built nave first, west to east.

Pointed, ribbed quadripartite vaulting (below) helped to further the aims of the builders. The vault consisted of a thin skin laid over strong ribs. Slender columns served to carry its weight.

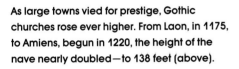

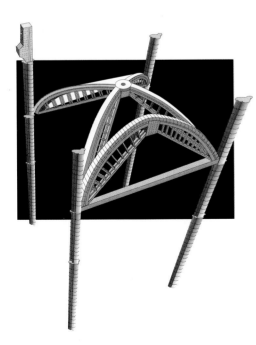

together brilliantly colored glass to fill the stone outlines of the enormous windows. In large earthen pots, the glassmakers mixed metal oxides with molten glass to obtain jewel-like colors: cobalt for blue, copper for red and green, and manganese for purple. The glass was often blown and worked into a cylindrical shape; then, once it had cooled, the glazier used a hot iron to cut it into the proper shapes and sizes—usually smaller than the palm of his hand. The details of the windows' sacred scenes were painted on with opaque enamel, which was then fired to fuse it completely to the glass. The whole translucent puzzle was supported by ironwork and held together with strips of lead.

Using simple hand tools (triangle and compass, plumb bob and square, metal ax and chisel, winch, carpenter's plane) and one new labor-saving device, the wheelbarrow, the cathedral builders realized the most complex industrial projects in Europe since Roman times. Their engineering advances were remarkable for the prescientific age. The organization of the work site itself—into coordinated teams of paid professionals, not slaves—remains perhaps their most modern achievement.

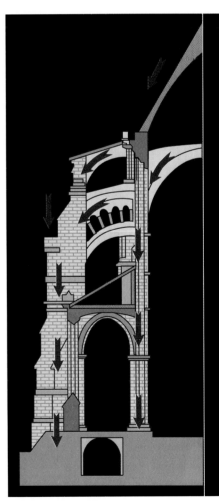

Flying buttresses, which bear against massive upright buttresses along the church's perimeter, restrain the outward thrust of the vaults and direct it downward, freeing the walls from carrying the load.

Gothic Cathedrals

Skeletal walls without flying buttresses support true curtain walls at Sainte-Chapelle, in Paris (opposite). Ever cautious, the builders reinforced the windows—just in case—with iron chains hidden in the tracery.

As technically advanced as the Gothic cathedrals were, structural efficiency does not seem to have been their builders' primary goal. For example, computer modeling reveals that the slender, steep, arching fliers on the buttresses at Bourges, completed in 1214, carry the building forces down to the ground in the most direct manner possible, with an almost modern minimum of structure. Yet the design was never repeated. In fact, 70 years later, in 1284, the immense choir at Beauvais collapsed—in part because the Bourges pattern of buttressing was not followed.

That the cathedrals still stand attests to the fact that the master masons had a highly developed understanding of structural principles. Yet all their engineering served to achieve one overriding goal: to become master illusionists. In this they succeeded: Their gigantic stone buildings seem not of this world.

Built in 1241 by Louis IX (St. Louis), little Sainte-Chapelle, in Paris, is visually perhaps the most perfect Gothic building. Because of its small scale, the engineering requirements for stability are far less than for the great cathedrals. Delicate looking quadripartite vaulting is supported by seemingly the thinnest of piers, without flying buttresses. The walls are not only pierced; they virtually disappear, hung with curtains of glass. The light sought by Abbot Suger of St. Denis comes pouring in, to the glory of God—and the glaziers.

Innovation continued to mark 13th-century building. By 1214, Bourges (upper) had the most efficient buttresses ever constructed.

The buttresses of Notre Dame de Paris (lower), first remodeled in the 1220s, represent a later but somewhat less efficient design.

WIND
SOLAR
HYDROELECTRIC

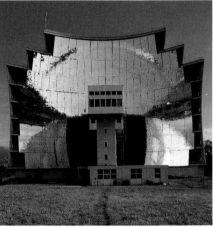

ODEILLO SOLAR FURNACE, FRANCE

THEODORE ROOSEVELT DAM, ARIZONA

OLD AND NEW WINDMILLS, CALIFORNIA

A lone in a landscape of pasture or polder, a windmill looks like a quaint anachronism—a poignant evocation of times past. Then, wind supplied a deceptively simple solution, mechanical power, to the problems of pumping water and grinding grain. The modern windmill is a wind turbine, which converts wind energy into electricity.

The search for renewable sources of energy that meet the world's needs without harming the environment is among the modern world's most compelling challenges. Most of the energy we use today still comes from highly polluting, non-renewable fossil fuels. Nuclear energy is also a resource, but nuclear power plants generate radioactive wastes, and even the safest of them carry a risk of disaster.

In addition to wind, the most promising alternative energy sources are water and the most reliable and renewable of all—the energy of the sun. In all three cases, the resource must be readily available; thus, siting is important. Wind turbines are most productive in windy areas; solar thermal plants work best in sunny, semiarid regions; hydroelectric dams are practicable only where water is plentiful.

The economical operation of wind turbines is possible only in areas that have annual average winds of at least 12 miles an hour. Although California generates most of the world's wind power, more wind

PRECEDING PAGES: Dozens of wind turbines re-
volve like pinwheels at California's Tehachapi Pass, some 75 miles north of Los Angeles. Strategically sited to catch the cool air currents that funnel through the pass into the Mojave Desert, the machines are part of a vast array of some 5,000 wind turbines at Tehachapi and more than 16,000 throughout the state. California now provides 75 percent of the world's wind-generated electricity, in the process supplanting 2.8 billion pounds of carbon dioxide and 16 million pounds of other pollutants that might otherwise be generated every year by power plants that run on coal and natural gas. New-generation wind turbines are being developed to compete with fossil fuels in the cost of producing electricity.

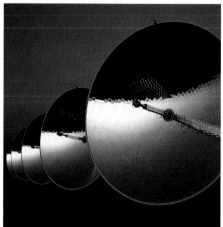

SOLAR ENERGY REFLECTORS, AUSTRALIA

actually blows over the strip of states from North Dakota to Texas—thus giving them even greater potential wind resources.

The blades of a wind turbine use aerodynamic lift to convert the kinetic energy of the wind into rotary motion. The two- or three-bladed turbine drives a generator. The generated electricity, its voltage boosted by transformers, flows along distribution lines into the local utility grid.

Wind turbines do not pollute. Also, in many areas wind-power costs are becoming competitive with costs of energy from fossil-fuel-burning plants. Wind turbines, however, are not entirely problem-free. They can clutter the landscape, and they can be noisy.

Solar power is generated by two basic types of energy-conversion systems: solar thermal and photovoltaic. In solar thermal systems, the sun's rays are focused onto receiving units by reflective parabolic dishes, troughs, or heliostats. Photovoltaic systems, on the other hand, rely on solar cells, which are made of semiconducting materials such as silicon.

Turbines and generators play a role in the production of solar thermal power. In this case, the intense heat generated in receiving units is used to make superheated steam, which can power a turbogenerator. Photovoltaic systems forgo turbines and generators and rely instead on solar cells to convert sunlight directly into electricity.

A hydroelectric facility also utilizes turbines, generators, and transformers, with a dam as the centerpiece. The dam—whether an embankment type built of earth or rock fill, or one made of concrete—stores water to power an electrical generator. Huge intake pipes, called penstocks, funnel water from the reservoir into the powerhouse, where it spins the turbines and their generators.

For all their promise, alternative energy systems are not without problems, most notably the need for practical yet inexpensive ways to store electricity for times when power plants are idled by lack of wind, sunshine, or water. Even so, the benefits of these systems are enormous: a lessened dependence on fossil fuels and a reduced threat to the environment.

LA GRANDE COMPLEX DAM, QUEBEC, CANADA

Wind

Modern-day counterparts of the old-fashioned windmill, Danish-built wind turbines sweep halos on the horizon in California's Altamont Pass.

Built in 1806 and still standing at East Hampton, New York, Hook Windmill sports sails spanning 60 feet, which project from a rotating "cap."

Wind is actually a form of solar energy, the result of the uneven heating of the earth's surface and the effects of its rotation. Wind is also an energy source in itself.

The earliest recorded attempt to snare the wind and tap its power by means of a windmill dates to the seventh century A.D. in Persia. By the 13th century, the technology had made its way to Europe, where two kinds of windmills eventually emerged: a post mill whose sails, attached to a horizontal beam, pivoted on a central vertical post to face the wind, and a tower mill whose sails were mounted on a rotating "cap." Such models were brought to North America and adapted. By the late 19th century there were as many as 6.5 million windmills in the United States, most of them grinding grain or pumping water.

More recently, the windmill, which transforms wind energy into mechanical power, has evolved into the wind turbine, which uses the wind to generate electricity. Today, more than 18,000 wind turbines operate in the United States, the vast majority in California, where three-fourths of the world's wind generated electricity is produced. The state's "wind farms" now supply enough power to meet all the residential needs of a city twice San Francisco's size, at a low cost to the consumer and the environment. Utility companies worldwide are showing interest in California wind farms.

Wind

Row upon row of horizontal-axis wind turbines, designed and manufactured by KENETECH/U.S. Windpower, stud the hillsides of an Altamont Pass wind farm. In 1991, the 80-foot towers generated 800 million kilowatt-hours, enough electricity for 130,000 average homes.

To scientists "prospecting" for wind resources in the early 1980s, California's Altamont Pass seemed an obvious choice for the site of what is today the world's largest concentration of wind turbines.

Indeed, as the mountain gap funnels wind into the hot Central Valley some 30 miles east of San Francisco, it also increases its speed—an important factor in turbine siting, because a little more wind speed translates into a lot more power. Wind power increases as the cube of wind velocity. Hence, a doubling of wind speed produces eight times as much power.

Even so, early attempts to harvest the wind at Altamont Pass were hardly encouraging. Poorly designed and inefficient, the first wind turbines had blades that sometimes flew off from the hub and generators that were prone to burnout. Altamont learned its lessons, becoming a proving ground for wind technology. Today, more than 7,000 wind turbines of various shapes and sizes are installed at the site, each one with a capacity to produce from 40 to 750 kilowatts of electricity.

Most of these machines are the familiar horizontal-axis, propellerlike wind turbines known as HAWTs, their high towers topped by turbine-generator units that support two or three fiberglass rotor blades up to 149 feet in diameter. These blades work in much the same way as the wings of an airplane: Air passing over the surface of the blades creates lift, which rotates the blades, spinning the shaft and driving the generator. Wires carry the generated electricity to a transformer in the tower's base, then to a substation and into the utility grid.

But more than 150 of Altamont's turbines owe their inspiration to a French engineer named G.J.M. Darrieus, who in 1931 invented a vertical-axis, "eggbeater" wind turbine, or VAWT, with a number of advantages over its horizontal-axis cousins. Unlike HAWTs, which rely on computer-controlled mechanical devices to keep their blades oriented into the wind, the long, curved blades of a Darrieus machine can capture wind from any direction.

Advances in gearbox and controller technology now allow horizontal-axis machines to operate with variable-speed rotors, extracting up to 10 percent more of the wind's power and creating less stress on the over-speed control system. These new, more efficient and reliable machines are also making wind technology cost-competitive with conventional coal-fired power plants and have led some experts to predict that within the next 30 years as much as 25 percent of U.S. power needs could be met by wind energy.

Americans are not unique in pursuing wind power. In Europe there is also considerable interest: Two German states alone plan to develop the capability to produce 2,000 megawatts of wind-generated electricity by the year 2010—compared to California's current capacity of 1,600 megawatts.

Altamont's vertical-axis "eggbeater" turbines (opposite) harvest wind from any direction.

Among the world's largest wind turbines is the computerized, 135-foot-tall model (above) atop Burgar Hill in the Orkney Islands of Scotland. Its 180-foot-diameter blades—almost as wide as the wingspan of a jumbo jet—can gather enough wind to generate 3,000 kilowatts of electricity.

Solar

At France's Odeillo Solar Furnace, large mirrors called heliostats collect sunlight and reflect it onto a parabolic mirror.

Although it is 93 million miles away, the sun is earth's single most important source of energy. In fact, the solar energy that makes its way to earth in one minute is more than the total amount of energy used by the planet's entire population in one year. The problem, of course, is that the sun's energy must be collected, concentrated, stored, and converted into other, more usable forms of energy.

Collecting and concentrating the sun's rays is as simple as holding a magnifying lens to the sun. Converting that radiation into other forms of energy is far more difficult. It wasn't until the late 19th century that scientists in Europe successfully converted sunlight into electricity. In this century, those early experiments have led to the development of two primary types of solar energy systems: photovoltaic systems, in which sunlight is converted directly into electricity by means of solar cells, and solar thermal systems, which convert sunlight into heat energy.

CALIFORNIA'S SOLAR ONE

Solar

The photovoltaic effect was first recognized in 1839, when a French physicist discovered that certain materials produce an electric current when light strikes them. This realization led to the development of photovoltaic, or solar, cells.

Solar cells are made from purified silicon and other semiconducting materials. These have different electronic properties and produce electric fields where they meet. As a cell absorbs light, electrons in the materials are released. Then the electric field drives them through an external circuit, producing a current.

Clouds and sky are mirrored in the reflector panels that bracket each bank of solar cells on California's Carrizo Plain. Both the cells and their reflector panels are mounted on a two-axis tracker designed to keep the photovoltaic array precisely oriented toward the sun.

Primitive devices, the first cells converted light to electricity with only a one percent efficiency rate. Today's commercial cells have efficiency rates as great as 14 percent. Laboratory rates are twice that.

Cells are usually grouped into modules, which are then assembled into arrays The simplest array uses flat-plate modules. Typically, these have a transparent cover, front or back structural supports, and layers of laminates to encapsulate the solar cells.

Photovoltaic systems are quiet, require no fuel, and generate no pollution. They are lightweight and portable. The U.S. space program first used one in 1958 to power a satellite's radio. Today, small systems power calculators and watches, while larger ones light highways and homes.

Some provide power to utility grids. The largest, on California's Carrizo Plain,

A manager at the Carrizo power plant checks a tracker's position on the computer monitor. A central computer ensures that each of Carrizo Plain's 756 trackers remains locked onto the constantly moving target of the sun.

Like futuristic sun-worshipers, hundreds of trackers bask in the California sunshine. Reflector panels further intensify the bombardment by photons that comprise sunlight, increasing the amount of electricity generated by each photovoltaic cell.

The 148-foot-high parabolic mirror of the Odeillo Solar Furnace, in southern France, works like an enormous magnifying lens to concentrate sunlight and focus it into a small area of intense heat. Almost 10,000 smaller mirrors make up the larger one. The furnace, which can melt almost any substance on earth, is located atop the tower in front of the mirror.

Solar

was built in 1984. Though since decommissioned, it supplied data for designing the next generation of large utility systems now in operation around the country.

Unlike photovoltaic systems, solar thermal systems do not generate electricity directly. They convert sunlight into heat energy, which can make steam to drive generators. But the heat is not always used to make steam. In fact, most thermal installations are solar water heaters.

Thermal systems use two basic types of solar collectors. The simplest, called flat-plate collectors, are used primarily to heat swimming pools. Water circulates through tubes bonded to black plates made of metal or nonmetallic compounds.

The other type is a concentrator collector, and it is mostly used commercially. It relies on differently shaped reflectors to increase heat conversion. These include U-shaped parabolic troughs, concave parabolic dishes, and large, flat mirrors called heliostats. All of them concentrate reflected sunlight onto receivers commonly filled with water, oil, or molten salt.

In facilities using a central-receiver configuration, heliostats focus sunlight onto a centrally located tower. Fluid in a receiver atop the tower is heated and transferred to an energy conversion unit in the tower's base. At Solar One, the world's largest such power plant when it was built in 1982, 1,818 heliostats focus on a 250-foot "power tower."

Solar radiation caroms from Odeillo's hillside heliostats to the huge parabolic mirror (diagram opposite), which focuses the light onto a receiver tower 58 feet away. Aluminum shutters at the tower's top open and close in less than a second, permitting a burst of light hotter than 6,500°F to penetrate the high-tech hearth.

The facility is being upgraded to Solar Two.

At least five southern California power plants use parabolic trough technology, with a total collector area of more than eight million square feet. Sunlight is concentrated onto a fluid-filled tube that extends along the axis of each trough.

Parabolic dishes reflect sunlight onto a receiver mounted above the center of each dish. As units track the sun, fluid in receivers is heated to more than 4,000°F. These versatile systems can operate autonomously or as part of larger systems.

A solar furnace can produce the highest temperatures. One of the largest was built in the 1960s, near Odeillo, France. It tested the heat-resistant tiles used on the U.S. space shuttles.

A state-of-the-art solar furnace at the National Renewable Energy Laboratory in Colorado concentrates the sun's energy up to 21,000 times. With a heating rate of a million degrees per second, the furnace could be used to make tough new alloys and might even be a means for destroying wastes generated by older technologies.

Sunrays burst from a tracking heliostat at the National Renewable Energy Laboratory in Colorado. Here, the world's newest solar furnace set a record for concentrating solar energy.

In a unique off-axis configuration, the tracking heliostat, at left, reflects sunlight onto a faceted primary concentrator. This device intensifies the light 2,500 times and focuses it onto a secondary concentrator, producing a beam 21,000 times the normal intensity of sunlight.

Hydroelectric

Rising more than 70 stories from bedrock, Glen Canyon Dam's concrete arch holds back the Colorado River in Arizona.

ROOSEVELT DAM, ON ARIZONA'S SALT RIVER.

ew structures are higher and none are heavier than dams, and among the highest and heaviest dams are those that trap earth's rivers to produce electricity.

The history of hydropower began in England in the mid-19th century with a pilot hydraulic project by engineer William Armstrong to electrify his Northumberland home. Large-scale electricity production had to wait until 1911, however, when the Theodore Roosevelt Dam, a 280-foot-high masonry-arch dam, opened in Arizona.

More ambitious projects followed, most notably the Hoover (formerly Boulder) Dam, completed in 1936, which harnessed the Colorado River, and Washington's Grand Coulee Dam, which used nearly 11 million cubic yards of concrete to control the Columbia River in 1942. One of the largest dam projects is Brazil and Paraguay's nearly five-mile-long Itaipú Dam on the Paraná River, opened in 1984 and capable of generating enough electricity to supply a city of three million people.

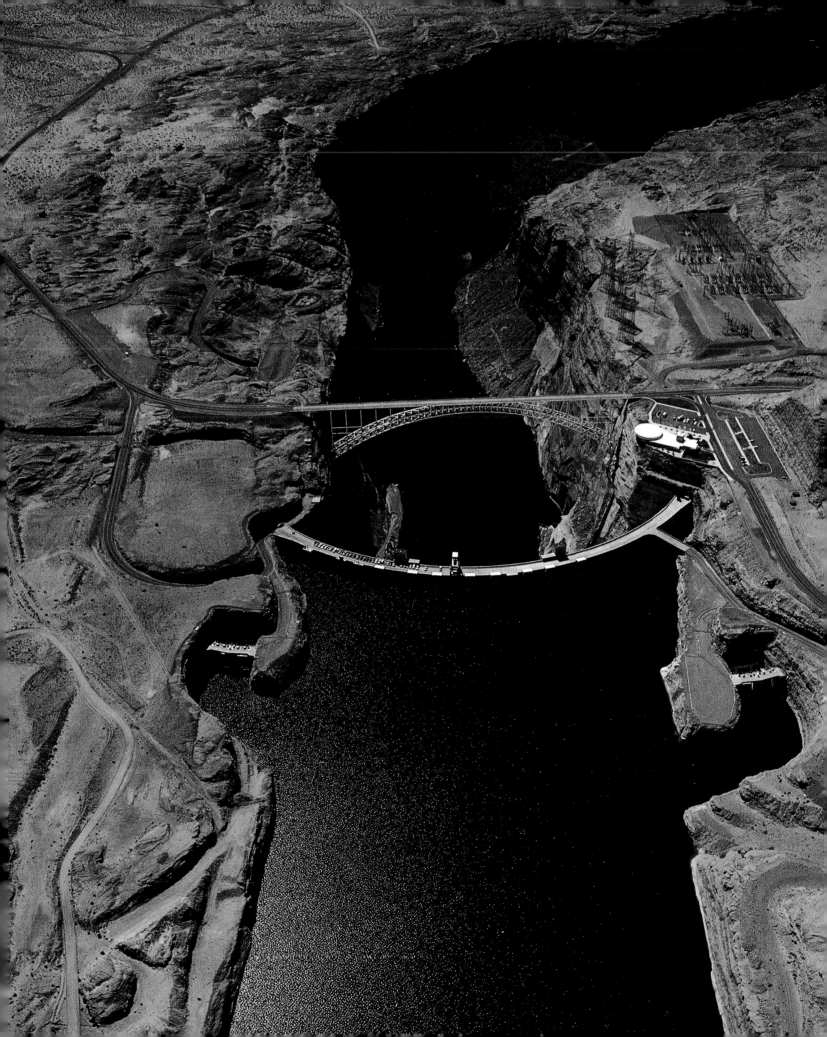

Building the Hoover Dam

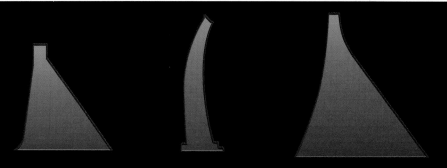

Although all concrete dams can contain rivers, they do so in different ways. A gravity dam (left) uses its own weight to withstand hydrostatic pressure—the horizontal thrust of water—and to channel it into the dam's foundation. An arch-gravity dam (center) relies on its shape to transfer the force of the water into the canyon walls. Some dams, including the Hoover (right), are hybrids, using both their weight and shape to resist the pressure.

With the dam's foundation excavated, workers swing the first of more than 400,000 buckets of concrete into position for pouring on June 6, 1933. Earlier, the Colorado River had been diverted through four 50-foot-high tunnels blasted into the canyon walls.

As a crew of "puddlers" (above, right) spreads a fresh layer of concrete in its wooden form, a worker uses a compressed-air vibrator to shake and settle the wet concrete.

he place was remote, miles from any highway or railroad, a desolate, rocky gorge on the Arizona-Nevada border— prone to high winds, sudden floods, and temperatures approaching 125°F. And yet to the engineers of the U.S. Bureau of Reclamation it was the ideal spot for what was to be then the world's largest dam.

Work on the Hoover Dam began in 1931 with the diversion of the Colorado River through four tunnels. Crews erected two cofferdams (temporary dams)—one upstream, the other downstream—to wall off the damsite from the river; then they pumped the site dry for excavation.

The excavating went on 24 hours a day until more than half a million cubic yards of mud and silt had been removed

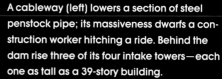

A cableway (left) lowers a section of steel penstock pipe; its massiveness dwarfs a construction worker hitching a ride. Behind the dam rise three of its four intake towers—each one as tall as a 39-story building.

The reinforced-concrete foundations of the dam's powerhouse (below) take shape in the foreground beside each of the canyon walls, as work reaches the halfway point in 1934.

John Lucian Savage (1879-1967) began his engineering career in 1903, after graduating from the University of Wisconsin and joining the U.S. Reclamation Service. Appointed the agency's chief design engineer in 1924, he supervised more than 60 dams, including the world's then largest concrete dams: Hoover, Grand Coulee, and California's Shasta. Many of the techniques pioneered in the Hoover's construction—among them the use of chilled water to speed the curing of concrete—can be attributed to Savage's ingenuity and are now routinely used in dam building. He later became a consultant on water projects worldwide.

from bedrock, and another million cubic yards of rock had been blasted from the canyon walls. Later, construction workers poured 6.6 million tons of concrete to create an arch-gravity dam nearly a quarter of a mile long and more than 70 stories tall.

All that concrete posed a problem for the engineers, however, since concrete, as it sets, gives off heat—so much in the case of the Hoover Dam that it would have taken more than a hundred years to completely cool. Moreover, as concrete cools, it shrinks and can crack. To speed up cooling and to prevent cracking, 582 miles of one-inch steel pipes were embedded in the wet concrete. Ice-cold water was then circulated through the pipes, permitting the concrete to set in just two years.

LG 2

Although western Canada is blessed with an abundance of oil and natural gas, the rest of the country is not so endowed. Instead, eastern Canada—the province of Quebec, in particular—has plenty of water, enough to drive one of the largest, and most controversial, hydroelectric facilities in the world.

Begun in 1971 and still under construction, La Grande Complex embraces some 68,000 square miles of northern Quebec near James Bay—an area as large as New England. Plans call for a series of nine dams to divert several rivers into La Grande River; operating at peak capacity, the dams' powerhouses will use the

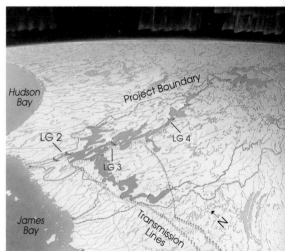

Crown jewel of La Grande Complex, LG 2 is part of an ambitious hydroelectric project designed to eventually cover some 68,000 square miles of eastern Canada.

More than 750,000 gallons of water per second cascade through LG 2's spillway and down a series of steps—a man-made cataract three times the height of Niagara Falls.

273

LG 2

harnessed water to generate 15,719 megawatts of electricity. The project continues to be held up, however, by environmental impact studies and by court actions on behalf of Native Americans, who fear its effects on their way of life.

Nevertheless, the first phase of the project was completed in 1985 and includes three colossal dams—LG 2, 3, and 4. Their three powerhouses together contain 37 turbines. LG 2, the largest of the dams, was a monumental undertaking by itself, requiring the excavation of some 3.3 million cubic yards of rock. Behind the dam, trillions of gallons of water backed up to form one of North America's biggest reservoirs. Spillover pours through the eight gates of LG 2's spillway and thunders down a man-made canyon designed to slow the water and prevent downstream erosion.

LG 2 is further distinguished by its

powerhouse, located 450 feet underground in a third-of-a-mile-long cavern blasted from solid rock. Turbines, fed by water carried from the reservoir, drive its 16 generators. Inside the powerhouse, two overhead, 400-ton traveling cranes ride on tracks that run the length of the central generator hall. The cranes helped install the generator units—whose rotors alone weigh 600 tons each—and remain to carry out routine maintenance and repairs.

Blasted from the ancient rock of the Canadian Shield, 37½-foot-high penstock tunnels (left) feed LG 2's turbines.

Some 450 feet below ground, LG 2's cavernous powerhouse area (opposite) took four years and 127,000 tons of explosives to excavate.

Central to LG 2's operation, mammoth turbines (below) have blades spun by water that plunges 590 feet through penstocks from the reservoir above.

The Itaipú Dam

S panning nearly five miles, containing 28 million tons of concrete, and employing 40,000 workers at the peak of construction, South America's Itaipú Dam is—in the estimation of the president of the company that built it—"the work of the century."

Few would disagree. Indeed, Itaipú,

An avalanche of water roars through the main spillway of the Itaipú Dam (left). To the right of the spillway is the hollow concrete main dam, its basic parts shown in the cross section below.

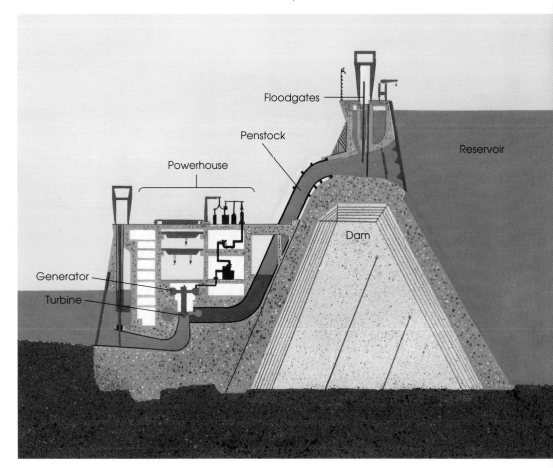

Floodgates

Penstock

Powerhouse

Reservoir

Generator

Turbine

Dam

which straddles the Paraná River along the border of Paraguay and Brazil, is one of the world's largest hydroelectric projects, its 18 generators capable of producing 12,600 megawatts of power. Not surprisingly, it was also one of the world's most expensive, costing some 18.3 billion dollars.

Construction began in 1974 with the excavation of a diversion channel to shunt the Paraná River around the proposed damsite. To build this 1.3-mile-long channel parallel to the river on the Brazilian side, workers over the next four years removed some 50 million tons of rock. They then constructed the first section of the main dam on the new channel, after building cofferdams near its entrance and exit to keep water out of the site. Once the first section was completed, in late 1978, the cofferdams were demolished, allowing the

The Itaipú Dam

river to flow through the diversion channel.

To ensure that the river stayed in its diversion channel while the rest of the main dam was under construction, crews erected cofferdams upstream and downstream of the main damsite and pumped dry the intervening area before excavating it to bedrock. Concrete was then poured, and the main dam and its power station slowly took shape; eventually, the dam rose to 643 feet. At the same time, work started on the spillway and on a buttress dam that connects the spillway to the main dam.

By the time Itaipú was structurally complete in 1984, so much concrete had been used—more than five times as much as went into the Hoover Dam—that three concrete-mixing plants had had to be built on-site. Even so, Itaipú was specifically designed to use as little concrete as possible. Indeed, unlike the arch-gravity Hoover Dam, Itaipú's main dam consists of hollow concrete segments grouted together to form a colossal wall with enough room inside to accommodate an enormous aircraft hangar. Moreover, most of Itaipú's great length is made up of two other kinds of dams that used little concrete, including more than three miles of embankment

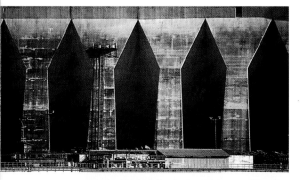

Massive concrete buttresses support a smaller, curved dam that links main dam to spillway.

At the heart of Itaipú lie the dam's 18 generators; together, they can produce 12,600 megawatts of electricity.

dams—composed of millions of tons of earth or rock fill—and the connecting dam, a rock-fill structure supported by massive concrete buttresses.

Itaipú's structural bulk is matched by its hydroelectric brawn. Each of the 18 generator units ensconced in the dam's cathedral-like powerhouse is fed by a 466-foot-long penstock. The generators themselves are the largest of their kind ever assembled, and each one measures 52 feet in diameter, stands 13 stories tall, and weighs an imposing 7,000 tons. Together, they provide enough electricity to meet all of Paraguay's present-day power needs and one-third of Brazil's.

Fourteen enormous steel gates (left) control access to Itaipú's quarter-mile-wide spillway. Three long chutes divide the spillway itself; down it, surplus reservoir water hurtles at 50 miles per hour (below).

FOLLOWING PAGES: Sheltered by the exterior wall of Itaipú's spillway, two workers appear indifferent to the explosion of water and mist around them. Some 10 million gallons of water can surge through the spillway every second.

G L O S S A R Y

Many technical terms used in *The Builders* are explained in the text and appear in the Index. This Glossary briefly defines important terms in the context of their usage in the book.

Anchorage: a device or object that provides a secure hold, such as the huge concrete block anchoring the main cable at each end of a suspension bridge.

Aqueduct: a conduit for conveying water downhill; a structure that carries a canal over a river or valley.

Arch: a structure, typically curved, that spans an opening and supports the weight above it.

Architrave: in classical architecture, the lowest part of an entablature.

Ashlar: squared and dressed stone; masonry having a uniform pattern of horizontal and vertical joints.

Batter: to incline from the vertical; for example, the walls of some medieval castles, which recede as they rise.

Beam: a horizontal supporting member of a structure.

Bluestone: a bluish-gray building stone.

Box girder: a beam that is a hollow rectangle when viewed end-on; lighter than a solid beam.

Buttress: a projecting structure, usually of masonry, that supports or stabilizes a building or wall.

Caisson: a watertight chamber used during underwater construction work and as a foundation for a bridge pier.

Campanile: a freestanding bell tower.

Cantilever: a horizontal structural member, such as a beam, girder, or truss, that projects from a supporting column or wall.

Capital: the top part of a column.

Cast iron: a mixture of iron, carbon, and silicon cast in a mold and used to make structural parts that are hard and nonmalleable but more fusible than steel.

Catenary arch: an arch having the same curved shape as that of a chain hanging freely from two fixed points—but turned upside down.

Centering: the temporary framework that supports a masonry arch while under construction.

Choir: the part of a church where services are sung.

Civil engineering: the designing and construction of public works such as dams, roads, canals, and bridges.

Cladding: a structure's outer skin that is attached to the frame after it is built.

Clerestory: a wall or story rising above an adjoining roof and containing windows.

Cofferdam: a watertight but open-to-the-air enclosure used primarily for excavating ground that is not deeply submerged.

Column: a vertical and usually round supporting pillar.

Compression: a force that shortens structural members by pressing or squeezing them together; opposite of tension.

Concrete: a hard building material made by mixing together a mineral aggregate (such as sand and gravel), a cementing material, and water.

Corbeling: successive masonry layers, each one projecting beyond the one immediately below it.

Cornice: a molded, horizontal projection that crowns the part to which it is attached.

Cramp: a device, usually of iron, that is used to pin timbers or stone blocks together.

Crenelation: a pattern of repeated dentations used, for example, in castle battlements.

Crossbeam: a beam that is set across another structural member.

Cross-bracing: a bracing system of intersecting diagonals.

Curtain wall: a non-load-bearing external wall that hangs from a skeleton frame.

Dead load: a constant load that results from a structure's weight, excluding occupants and contents.

Dome: usually a hemispherical ceiling or roof.

Dowel: a pin that fits into a hole in an abutting piece and prevents slippage.

Entablature: in classical architecture, a horizontal beam supported by columns.

Facing: a structure's outer masonry layer, used as ornamentation or for protection.

Flying buttress: a masonry structure that bears against an upright buttress on a church's external wall and directs downward the outward thrust of a vault.

Frame: the supporting structure, or skeleton, of a building.

Frieze: the part of an entablature between the cornice and the architrave.

Geodesic dome: a structure consisting of many light, straight elements, usually in tension, that form a grid shaped like a dome.

Girder: a large or principal beam in a structure.

Gnomon: an object positioned so that its shadow indicates the time of day.

Heading: the end of a horizontal, underground passageway or tunnel.

Heliostat: a mirrored instrument used to track the sun and reflect its light onto a receiver.

Hoop tension: the tension around the lower part of a dome.

Hypotenuse: the side opposite the right angle in a right-angled triangle.

Inclined plane: an inclined track on which heavy objects can be moved.

Joint: a place where two structural members are joined.

Keep: the stronghold of a castle.

Lintel: a short, horizontal beam that spans an opening.

Live load: the load beyond the weight of a structure itself; includes occupants and contents.

Load: the force placed on a structure by weight or wind pressure.

Load-bearing: capable of carrying a load in addition to its own weight.

Lock: a canal enclosure, with gates at both ends, used to lower or raise vessels from one level to another.

Macadamize: to construct a pavement by compacting layers of crushed stones on a convex roadbed and binding them together with cement or asphalt.

Machicolation: openings in a castle's parapet that are used for dropping missiles on enemies below.

Modular construction: construction that uses standardized dimensions or units.

Mortise: a cavity into which some other part, such as a tenon, is inserted.

Nave: the main part, or central aisle, of a church.

Oculus: something resembling an eye, especially a round opening at the crown of a dome.

Parabolic dish: a reflective structure that is dish- or bowl-shaped and focuses sunlight onto a receiver mounted above the center of the dish.

Parabolic trough: a reflective structure that has a curved, elongated shape and focuses sunlight onto a receiver running the length of the trough.

Pendentive: a triangular segment of vaulting that supports a circular dome over a square space.

Penstock: a conduit for water, especially one that controls the flow of water from a reservoir to a turbine.

Pentelic marble: marble that comes from Mount Pentelicus in Greece.

Photovoltaic cell: a device that is used to convert sunlight directly into electricity.

Pier: an upright structure, such as a column, a pillar, or a pilaster, that supports a vertical load.

Pilaster: a shallow, rectangular pier or column attached to a wall.

Portcullis: a retractable iron gate.

Precast: formed in the shape of a structural element, such as a concrete beam or panel, at a site other than where it will be used.

Prefabrication: factory production of standardized parts that can be assembled later at a construction site.

Prestressed concrete: concrete in which steel tendons have been introduced, stretched taut, and anchored at either end. The ends pull together and thus create an upward force that counterbalances the downward force of the applied loads.

Purlin: a horizontal timber supporting the rafters in a roof.

Quadrant: one-quarter of a circle, or a device shaped like a quarter of a circle.

Reinforced concrete: concrete that has metal rods, wires, or bars embedded in it to add strength.

Rivet: a metal bolt that is passed through holes in two metal plates and then hammered so that its plain end forms a second head and securely unites the plates.

Sarsen: a large sandstone block.

Semiconductor: a crystalline material, such as silicon, whose electrical conductivity is intermediate between an insulator and a metal.

Sill: a horizontal piece that is the lowest part of a framework or other supporting structure.

Slip form: a massive mold that is moved slowly up a structure, usually by hydraulic jacks, as concrete hardens within the mold.

Space frame: a three-dimensional structure composed of interconnected members in which stresses are equally distributed.

Span: the distance between supports in a structure.

Spillway: an opening or passage through which surplus water is released from a reservoir.

Stainless steel: a steel alloy that is highly resistant to rusting.

Steel: a commercial iron compound containing less carbon than cast iron and more than wrought iron, and which is highly malleable.

Stiffen: to increase a structure's resistance to deformation.

Stress: the force exerted as one body twists, pulls, presses, or pushes against another body; also, the intensity of this force as expressed in pounds per square inch.

Strut: a structural brace that resists pressure in the direction of its own length; can be vertical, horizontal, or diagonal.

Talus: a slope of soil and rock debris.

Tendon: in prestressed concrete, a steel band that is fitted into a concrete beam.

Tenon: a projection on a piece of wood or stone that is inserted into a mortise on another piece.

Tensile strength: the ability of a material to resist stretching forces.

Tension: a force that pulls outward on the members of a structure, stretching or lengthening them; opposite of compression.

Thrust: the amount of force exerted by or on a structure.

Tongue-and-groove joint: a joint made when the tongue-like part on the edge of one object is fitted into a groove on the edge of another object.

Transept: the transverse arms of a cruciform church, usually separating the nave and the choir.

Truss: a framework of structural members, such as beams or girders, that strengthen each other and together form a long beam.

Vault: an arched structure, usually of masonry, that forms a roof or ceiling.

Viaduct: a bridge or series of bridges carrying a roadway or railway over an obstacle such as a valley.

Voussoir: a wedge-shaped block of stone used in a masonry arch or vault.

Waves: undulatory or rolling movements, including vibrations caused by an earthquake, that pass through air, water, and the earth.

Wind load: the force on a structure, or part of a structure, caused by the wind.

Wind shear: the stress on a structure when winds of different directions or velocities are close together.

Wind turbine: a machine that converts the force of the wind into electrical energy.

Wrought iron: a form of iron having a lower carbon content than steel or cast iron but which is tough and malleable.

The Book Division gratefully acknowledges the generous help of many individuals in preparing this volume. Our special thanks go to Robert M. Vogel, retired curator of mechanical and civil engineering, Smithsonian Institution; William L. MacDonald; and David P. Billington and Robert Mark, Princeton University.

We would also like to thank Charles W. Boning, U.S. Geological Survey; Tim Brown, American Society of Civil Engineers; Pam Byer, CN Tower; John B. Carlson, Center for Archaeoastronomy; Scott Collins and Claude Wolff, Embassy of France; Craig Culp, American Wind Energy Association; Nathaniel Curtis, The Curtis Architectural Corporation; Darrell Dodge, Patrick Summers, and John Thornton, National Renewable Energy Laboratory; Neal Fitzsimons; Charles F. Gay and Barbara Isenburg, Siemens Solar Industries; Hermann Guenther of Daniel, Mann, Johnson & Mendenhall; James Harle, Alyeska Pipeline Service Company; John Hounslow, The Thames Barrier Operational Area; Folke T. Kihlstedt, Franklin and Marshall College; Mark Lehner, University of Chicago; Valerie Mattingley and Heather Yule, NGS-London Office; Kerri E. Miller, KENETECH/U.S. Windpower; Kathleen Moenster, Jefferson National Expansion Historical Association; Michael Nylan, Bryn Mawr College; David O'Connor, University of Pennsylvania; Joseph Passoneau, Joseph Passoneau & Partners; Alison Porter, Eurotunnel; Julian Rhinehart, U.S. Bureau of Reclamation; Julian Richards, AC Archaeology; Jeffrey Stine, William Withuhn, and William Worthington, Jr., Smithsonian Institution; Homer A. Thompson, Princeton University; J. van Duivendijk, Royal Dutch Consulting Engineers; Richard M. Vogel, Tufts University; Arthur Waldron, Naval War College.

At the National Geographic, we are indebted to the Pre-Press/Typographic Division, especially Jessica P. Norton, and to the Library and its News Collection, the Illustrations Library, and the Photographic Laboratory.

ILLUSTRATIONS CREDITS

Abbreviations for terms appear below: (t)-top; (b)-bottom; (l)-left; (r)-right; (c)-center; NGP-National Geographic Photographer; NGS-National Geographic Staff.

1, Jake Rajs. 2-3, Winfield Parks. 6-7, Cameron Davidson/COMSTOCK.

OVERCOMING DISTANCE: 8-9, Jay Maisel. 10, (l) William H. Clark/FPG, INT'L; (tr) Dan McCoy/ BLACK STAR; (br) Craig Aurness/WESTLIGHT. 11, (l) Gary J. Benson/COMSTOCK; (r) John Lawler/TSW. **ROADS:** 12, American Society of Civil Engineers. 13, Alex S. MacLean/PETER ARNOLD, INC. 14-15, Peter Essick. 15, (t) Dale Glasgow; (c) and (b) Dale Glasgow based on an illustration from: *PAST WORLDS: The Times Atlas of Archaeology*, published by Times Books, a division of HarperCollins Publishers. 16, (t) Robert Frerck/ODYSSEY-CHICAGO; (b) H.W. Silvester/PHOTO RESEARCHERS, INC. 16-17, (t) Cary Wolinsky. 17, Cary Wolinsky. 18-19, Kenneth Garrett. 19, ANN RONAN AT IMAGE SELECT. 20, (t) THE BETTMANN ARCHIVE; (b) G. Colliva/THE IMAGE BANK. 21, Ric Ergenbright. 22, (t) Nicholas DeVore III-PHOTOGRAPHERS/ASPEN, INC.; (b) Colorado Department of Transportation. 23, Courtesy, Joseph Passoneau. 24, (t) Nicholas DeVore III-PHOTOGRAPHERS/ASPEN, INC.; (b) Hermann Guenther. 25, (tl) Hermann Guenther; (tr) Nicholas DeVore III-PHOTOGRAPHERS/ASPEN, INC.; (b) Colorado Department of Transportation. **CANALS:** 26, DEUTSCHES MUSEUM, MÜNCHEN. 27, Ric Ergenbright. 28, Georg Gerster. 29, (l) Ric Ergenbright; (r) Dean Conger; (b) Dale Glasgow. 30, (l) ANN RONAN AT IMAGE SELECT; (r) Linda Bartlett. 31, (t) ANN RONAN AT IMAGE SELECT; (b) Linda Bartlett. 32, THE GRANGER COLLECTION, New York. 32-33, Bob Sacha. 34, (t) Dale Glasgow; (b) Erich Hartmann/MAGNUM. 35, (all) NEW YORK PUBLIC LIBRARY; 36-37, Robert Frerck/ODYSSEY-CHICAGO. 37, Will & Deni McIntyre/PHOTO RESEARCHERS, INC. 38, (tl) THE GRANGER COLLECTION, New York; (bl) George F. Mobley, NGP; (r) THE BETTMANN ARCHIVE. 39, (tl) THE GRANGER COLLECTION, New York; (bl) THE BETTMANN ARCHIVE; (tr) FROM THE COLLECTIONS OF THE LIBRARY OF CONGRESS; (br) George F. Mobley, NGP. 40, (t) From "The Inauguration of the Suez Canal," by Marius Fontaine; Illustration by M. Riou. Photographed by Jonathan S. Blair; (b) Thomas J. Abercrombie, NGP. 41, Dallas & John Heaton, STOCK.BOSTON. 42, Michael S. Yamashita. 42-43, Robert Frerck/ODYSSEY-CHICAGO. 43, (t) and (b) DeFoy/MAURITIUS. 44, (t) SEFTON PHOTO LIBRARY, MANCHESTER; (b) Georg Gerster/COMSTOCK. 45, (l) Georg Gerster; (r) Blaine Harrington. 46-47, Blaine Harrington. 48, (l) Georg Gerster/COMSTOCK; (r) James P. Blair, NGP. 49, (l) "Saqiya Water-Lifting Device" Artist, George Resteck from "Mechanical Engineering in the Medieval Near East," Donald R. Hill. Copyright © 1992 by Scientific American, Inc. All rights reserved; (tr) and (br) Thomas J. Abercrombie, NGP. **BRIDGES:** 50, Dale Glasgow. 51, Andy Levin/PHOTO RESEARCHERS, INC. 52, (tl) Daryl Benson/MASTERFILE; (bl) Dean Conger; (r) Porterfield & Chickering/PHOTO RESEARCHERS, INC. 53, (t) Thomas Kitchin/TOM STACK & ASSOCIATES; (b) Lowell Georgia/PHOTO RESEARCHERS, INC. 54, (t) Dean Conger; (b) SEFTON PHOTO LIBRARY, MANCHESTER. 55, Museum/Photo: Michael Holford. 56, (tl) and (bl) COLLECTIONS/ Brian Shuel; (r) THE GRANGER COLLECTION, New York. 57, Michael Holford. 58-59, Graeme Outerbridge. 59, Mancia/Bodmer, FBM Studio, Switzerland. 60, (bl) Oliver Pighetti/GAMMA LIAISON; (br) Loren McIntyre. 60-61, J.A. Kraulis/MASTERFILE. 62, (b) THE BETTMANN ARCHIVE. 62-63, Hiroyuki Matsumoto/BLACK STAR. 63, (b) MUSEUM OF THE CITY OF NEW YORK. 64, (t) BROWN BROTHERS; (c) Dale Glasgow; (b) THE GRANGER COLLECTION, New York. 65, Ken Straiton/THE STOCK MARKET. 66, 67 (all) Gabriel Moulin. 68, Robert Van Marter/GAMMA LIAISON. 69, (t) Jake Rajs; (b) Torin Boyd. 70, (all) SMITHSONIAN INSTITUTION. 71, (t) Graeme Outerbridge; (b) Leslie Garland. 72, (t) Graeme Outerbridge; (b) Adam Wolfitt/SUSAN GRIGGS AGENCY. 73, Graeme Outerbridge. 74-75, Cameron Davidson/BRUCE COLEMAN INC. 75, (all) Graeme Outerbridge. **RAILROADS:** 76, Courtesy, THE OAKLAND MUSEUM HISTORY DEPARTMENT. 77, Gary J. Benson/COMSTOCK. 78-79, (all) Courtesy, THE OAKLAND MUSEUM HISTORY DEPARTMENT. 80-81, DENVER PUBLIC LIBRARY. 81, Lowell Georgia. 82, Gary J. Benson/COMSTOCK. 82-83, Raga/THE STOCK MARKET. **PIPELINES:** 84, (all) FROM THE COLLECTIONS OF THE LIBRARY OF CONGRESS. 85, W.R. Moore. 86, John Lawlor/TSW. 87, (tl) and (cl) Steve Raymer, NGS; (bl) Dale Glasgow; (r) Lloyd Sutton/MASTERFILE. 88, (inset) Thomas Kitchin/TOM STACK & ASSOCIATES. 88-89, Richard Schlecht.

HEIGHT AND DEPTH: 90-91, Jake Rajs. 92, (l) and (tr) Jake Rajs; (br) Raphael Gaillarde/GAMMA LIAISON. 93, (l) Jodi Cobb, NGP; (r) George Hall/ WOODFIN CAMP & ASSOCIATES. **TOWERS:** 94, (t) ANN RONAN AT IMAGE SELECT; (b) CULVER PICTURES, INC. 95, Jay Maisel. 96, Jake Rajs. 97, Gordon Gahan. 98, (l, all) THE BETTMAN ARCHIVE; (tr) Marc Riboud/ MAGNUM. 98-99, Miguel/THE IMAGE BANK. 99, (l) THE BETTMANN ARCHIVE; (tr) and (br) ANN RONAN AT IMAGE SELECT. 100-101, Georg Stärk. 102, (l) Lance Nelson/SUSAN GRIGGS AGENCY; (r) A.C. Chinn/Washington Evening Star. 103, (tl) Source unknown; (b) NATIONAL PARK SERVICE; (tr) Earl Young/ROBERT HARDING PICTURE LIBRARY-overlay by Dale Glasgow. 104, Kunio Owaki/THE STOCK MARKET. 105, Charlie Palek/TOM STACK & ASSOCIATES. 106, (l) George F. Mobley, NGP; (tr) Bruce Dale, NGP; (br) James P. Blair, NGP. 107, (t) Arthur L. Witman; (b) Dale Glasgow. 108, From *Structures* by Nigel Hawkes, Marshall Editions Ltd. 109, Bill Brooks/MASTERFILE. 110, (l) STATOIL; (r) Dale Glasgow. 111, STATOIL. **TUNNELS:** 112, ANN RONAN AT IMAGE SELECT. 113, Raphael Gaillarde/GAMMA LIAISON. 114, THE ILLUSTRATED LONDON NEWS PICTURE LIBRARY. 114-115, 115, ANN RONAN AT IMAGE SELECT. 116, (l) CULVER PICTURES, INC. (r) THE GRANGER COLLECTION, New York. 117, (t) Heinrich Berann; (b) D. Beal/BLACK STAR. 118, 118-119, (both) ANN RONAN AT IMAGE SELECT. 120, (b) Paul Van Riel/BLACK STAR. 120-121, From *Structures* by Nigel Hawkes, Marshall Editions Ltd. 121, (t) Photoworld/FPG, INT'L; (b) FROM THE COLLECTIONS OF THE LIBRARY OF CONGRESS. 122-123, Michael Koester/ARTWORKS. 123, (inset) Source unknown. 124, (t) Steve Drexler/THE IMAGE BANK; (b) QA PHOTOS. 125, (tl) Raphael Gaillarde/GAMMA LIAISON; (tr) N'Diaye /ARCHIVE PHOTOS; (b) From *How They Were Built* by David Brown. Published by Kingfisher Books. Copyright © Griswood & Dempsey Ltd. 1991. **SKYSCRAPERS:** 126, (t) THE BETTMANN ARCHIVE; (b) BROWN BROTHERS. 127, Masa Uemura/ALLSTOCK. 128, Hedrich-Blessing Photo Courtesy CHICAGO HISTORICAL SOCIETY. 128-129, William H. Bond, NGS. 129, (t) John Lewis Stage/THE IMAGE BANK; (b) Nathan Benn. 130, Norman Thomas and W.W. Rock. 131, (tl) Leah Schnall; (bl) FROM THE COLLECTIONS OF THE LIBRARY OF CONGRESS; (r) THE BETTMANN ARCHIVES. 132, (l) Medsciartco/Tomo; (r) *Science & Building* by Henry J. Cowan. 133, Jake Rajs. 134, (l) Thomas Laird/PETER ARNOLD, INC.; (r) Lewis Hine, Courtesy International Museum of Photography at George Eastman House. 135, (all) Lewis Hine, Courtesy International Museum of Photography at George Eastman House. 136, Alex Quesada/MATRIX. 137, (l) Nick Merrick, Hedrich-Blessing; Photo courtesy of architects Skidmore, Owings & Merrill; (r) Lynn Johnson. 138-139, Wernher Krutein/GAMMA LIAISON. 139, (l) Courtesy Skidmore, Owings & Merrill; (r) Bill Brooks/MASTERFILE. 140, (t) LEBARON COLLECTION/John LeBaron; (b) Jake Rajs. 141, (t) Dale Glasgow; (b) James A. Sugar. 142-143, James Marshall/THE STOCK MARKET. 144, (t) Lovell/MAURITIUS; (c) and (b, both) Ove Arup & Partners. 145, (l) G.V. Faint/THE IMAGE BANK; (r) Paul Warchol.

PUBLIC SPACES: 146-147, Michael S. Yamashita. 148, (l) Stumpf/SIPA IMAGE; (r) C.M. Dixon. 149, (tl) Joseph H. Bailey, NGP; (bl) E. Tavora de Santo/SUPERSTOCK; (r) L.J. MacDougal/COMSTOCK.CANADA. **SPORTS ARENAS:** 150, THE IMAGE BANK. 151, Alan Becker/THE IMAGE BANK. 152-153, Richard Steedman/THE STOCK MARKET. 154, (t) Quarto Publishing PLC; (b) Georg Gerster/COMSTOCK. 155, (t) Richard Steedman/THE STOCK MARKET; (b) Quarto Publishing PLC. 156, John Riley/FOLIO, INC. 157, (t) From *Structures* by Nigel Hawkes, Marshall Editions Ltd.; (b) Nathaniel Curtis, F.A.I.A. Architect. 158, (t) Nathaniel Curtis, F.A.I.A. Architect. 158-159, (b) Peter Frey/THE IMAGE BANK. 159, (t) Nathaniel Curtis, F.A.I.A. Architect. 160, (t) Horst Schafer/PETER ARNOLD, INC.; (b) Ingrid Otto. 161, (t) Jurgen Schadeberg/SUSAN GRIGGS AGENCY; (b) Hetz/MAURITIUS. **EXPOSITION HALLS:** 162, THE GRANGER COLLECTION, New York. 163, Bill Brooks/MASTERFILE © The Walt Disney Company. 164, (t) ANN RONAN AT IMAGE SELECT; (b) THE GRANGER COLLECTION, New York. 165, (l) THE GRANGER COLLECTION, New York; (r) THE ILLUSTRATED LONDON NEWS PICTURE LIBRARY. 166, 167, (all) ANN RONAN AT IMAGE SELECT. 168, J.A. Kraulis/ MASTERFILE. 168-169, Tommy Thompson/ BLACK STAR. 170, Jay Maisel. 171, (t) Mark Segal/ODYSSEY-CHICAGO; (b) Everett C. Johnson/FOLIO, INC. 172-173, S. Machado/SUPERSTOCK. 173, (t) Joachim Messerschmidt/WESTLIGHT; (b) Randy Taylor/ GAMMA LIAISON.

THE NEED FOR PROTECTION: 174-175, Tor Eigeland. 176, (l) Michael St. Mauer Sheil/SUSAN GRIGGS AGENCY; (r) Adam Woolfitt/SUSAN GRIGGS AGENCY. 177, (l) M. Keller/SUPERSTOCK; (r) Michael S. Yamashita/WESTLIGHT. **ON LAND:** 178, *The World Atlas of Architecture*, published by Mitchell Beazley International Ltd. 179, Dean Conger. 180-181, From *Structures* by Nigel Hawkes, Marshall Editions Ltd. 181, Hubertus Kanuf/SUPERSTOCK. 182, Dale Glasgow. 182-183, Adam Woolfitt. 183, James L. Stanfield, NGP. 184, Dale Glasgow. 184-185, Georg Gerster/COMSTOCK. 186, (l) The drawing is taken from (translating from the Chinese) *The History of Ancient Chinese Architecture*. Beijing. 1980. (r) Dean Conger. 187, Dallas and John Heaton/STOCK.BOSTON. 188, William Albert Allard. 189, (l) Ric Ergenbright; (r) J.C. Carton/ BRUCE COLEMAN INC. **FROM WATER:** 190, Pablo Bartholomew. 191, Ovak Arslanian/GAMMA LIAISON. 192-193, Dean Conger. 193, Weller Fishback & Bohl Architects. 194-195, O. Louis Mazzatenta, NGS. 195, (t) COLLECTIONS/Brian Shuel; (c) O. Louis Mazzatenta, NGS; (b) Dale Glasgow. 196, Bart Hofmeester/AEROCAMERA. 197, (tl) Patrick Ward; (tr) and (bl) and (br) Ovak Arslanian. 198, (l) and (r) William H. Bond, NGS. 199, Michael St. Maur Sheil. 200, 201, (all) Pablo Bartholomew.

RESPONDING TO THE SPIRIT: 202-203, Jay Maisel. 204, (l) P. & R. Manley/SUPERSTOCK; (r) Robert Frerck/ ODYSSEY-CHICAGO. 205, (tl) HIRMER FOTOARCHIV MÜNCHEN; (bl) C.M. Dixon; (r) Lawrence Migdale/ PHOTO RESEARCHERS, INC. **PYRAMIDS:** 206, From *Description de l'Egypte*, France, Commission des Monuments d'Egypte, Paris, 1809-1828. 207, Fred Maroon. 208, (l) Dale Glasgow; (r) CULVER PICTURES, INC. 209, (l) Roger Wood/PICTUREPOINT; (r, all) Fred Maroon. 210-211, John G. Ross/PHOTO RESEARCHERS, INC.; 211, From "HOW IS IT DONE?" © Reader's Digest 1990, illustration by Gerald Eveno. 212, (l, both) Dale Glasgow; (r) NGS Collection. 213, Will & Deni McIntyre/PHOTO RESEARCHERS, INC. **TEMPLES:** 214, Georg Gerster. 215, E. Streichan/SUPERSTOCK. 216, (both) Dale Glasgow. 216-217, B. VanBerg/THE IMAGE BANK. 217, Dale Glasgow. 218, (t) Dean Conger; (b) Weller, Fishback & Bohl. 219, (tl) Dean Conger; (tr) Weller, Fishback & Bohl; (b) Vidler/MAURITIUS. 220, 221, Dean Conger. 222-223, From *How They Were Built* by David Brown. Published by Kingfisher Books. Copyright © Grisewood & Dempsey Ltd. 1991. 224, (t) Otis Imboden; (b) SEFTON PHOTO LIBRARY, MANCHESTER. 225, (l) Harald Sund/THE IMAGE BANK; (tr) V. Wilkinson/VALAN PHOTOS; (br) TravelPix/FPG INT'L. **DOMES:** 226, Dale Glasgow. 227, Stuart N. Dee/THE IMAGE BANK. 228-229, From *How They Were Built* by David Brown. Published by Kingfisher Books. Copyright © Grisewood & Dempsey Ltd. 1991. 230, James L. Stanfield, NGP. 231, K. Scholtz/SUPERSTOCK. 232, Dale Glasgow. 232-233, From *The World Atlas of Architecture*, published by Mitchell Beazley International Ltd. 234, "A Proposed View of Brunelleschi's Machines in Operation." Artist, George Resteck from "Building the Cathedral in Florence" by Gustina Scaglia. Copyright © 1992 by Scientific American, Inc. All rights reserved. 235, Michael S. Yamashita. 236-237, James L. Stanfield, NGP. 237, ART RESOURCE/SCALA. 238, (l) Dale Glasgow; (tr) BROWN BROTHERS; (b) Larry Fisher/MASTERFILE. 239, From *The World Atlas of Architecture*, published by Mitchell Beazley International Ltd. 240-241, Jay Dickman. **GOTHIC CATHEDRALS:** 242, (l) Aime Roux/EXPLORER; (r) Louis Salou/EXPLORER. 243, Adam Woolfitt/WOODFIN CAMP & ASSOCIATES. 244, Ken Straiton/THE STOCK MARKET. 245, (l) Joseph Nettis/PHOTO RESEARCHERS, INC.; (r) Bruce Dale, NGP. 246-247, Hartman-DeWitt/COMSTOCK. 248, (l) Harry Bliss; (r, both) Dale Glasgow. 249, THE GRANGER COLLECTION, New York. 250, Michael Holford. 250-251, (t) Robert Mark, Princeton University. 251, (bl) and (br) Dale Glasgow. 252, (t) Chor von Osten/HIRMER FOTOARCHIV MÜNCHEN; (b) Ted Mahieu/THE STOCK MARKET. 253, James L. Stanfield, NGP.

HARNESSING NATURE'S POWER: 254-255, Ron Sanford/BLACK STAR. 256, (l) David Frazier/SIPA PRESS; (tr) E. Gebhardt/MAURITIUS; (br) Walter H. Hodge/PETER ARNOLD, INC. 257, (l) Otto Rogge/THE STOCK MARKET; (r) Renaud Thomas/FPG INT'L. **WIND:** 258, FROM THE COLLECTIONS OF THE LIBRARY OF CONGRESS. 259, Wernher Krutein/LIAISON INT'L. 260, (t) Photo Courtesy of KENETECH/U.S. Windpower, Inc.; (b) James A. Sugar/BLACK STAR. 261, Michael Marten/Science Photo Library/PHOTO RESEARCHERS, INC. **SOLAR:** 262, Harald Sund/THE IMAGE BANK. 263, Georg Gerster/COMSTOCK. 264-265, Harald Sund/THE IMAGE BANK. 265, (t) and (b) Jean Marc Giboux/GAMMA LIAISON. 266, (t) Emory Kristof, NGP; (b) From *Structures* by Nigel Hawkes, Marshall Editions Ltd. 267, (r) and (b) National Renewable Energy Laboratory/U.S. Department of Energy. **HYDROELECTRIC:** 268, Stan Osolinski/FPG, INT'L. 269, Dewitt Jones. 270, (t) Dale Glasgow; (bl) and (br) FROM THE COLLECTIONS OF THE LIBRARY OF CONGRESS. 271, (tl) and (bl) FROM THE COLLECTIONS OF THE LIBRARY OF CONGRESS; (r) Photoworld/FPG INT'L. 272-273 Ottmar Bierwagen. 273, Lloyd K. Townsend. 274, (t) George Loehr/THE IMAGE BANK; (b) Peter Christopher/MASTERFILE. 275, Peter Christopher/MASTERFILE. 276-277, Camara Tres/MAURITIUS. 277, Illustration by Dean Ellis; Reprinted from *Popular Mechanics*, July 1985. © Copyright The Hearst Corporation. All Rights Reserved. 278, (t) Daniel Aubry; (b) Wouterloot-Gregoire/VALAN PHOTOS. 279, (t) Daniel Aubry; (b) Frederico Mendes/SIPA PRESS. 280-281, Randa Bishop.

Boldface indicates illustrations.

Library of Congress CIP Data
The Builders : marvels of engineering / prepared by the Book Division, National Geographic Society.
p. cm.
Includes index.
ISBN 0-87044-836-6. -- ISBN 0-87044-837-4 (Deluxe)
1. Engineering--History. I. National Geographic Society (U.S.). Book Division.
TA15.B85 1992   92-30615
620--dc20   CIP

Composition of this book by the Typographic section of National Geographic Production Services, Pre-Press Division. Color separations by Graphic Art Service, Inc., Nashville, Tenn.; Lanman Progressive Co., Washington, D.C.; Lincoln Graphics, Inc., Cherry Hill, N.J.; and Phototype Color Graphics, Pennsauken, N.J. Printed and bound by Arcata Graphics/Kingsport, Kingsport, Tenn. Paper by Mead Paper Co., New York, N.Y. Dust jacket printed by Arcata Graphics/Kingsport, Kingsport, Tenn.